Geoheritage, Geoparks and Geotourism

Conservation and Management Series

Spectacular geo-morphological landscapes and regions with special geological features or mining sites are becoming increasingly recognized as critical areas to protect and conserve for the unique geoscientific aspects they represent and as places to enjoy and learn about the science and history of our planet. More and more national and international stakeholders are engaged in projects related to "Geoheritage", "Geo-conservation", "Geoparks" and "Geotourism"; and are positively influencing the general perception of modern Earth Sciences. Most notably, "Geoparks" have proven to be excellent tools to educate the public about Earth Sciences; and they are also important areas for recreation and significant sustainable economic development through geotourism. In order to develop further the understanding of Earth Sciences in general and to elucidate the importance of Earth Sciences for Society, the "Geoheritage, Geoparks and Geotourism Conservation and Management Series" has been launched together with its sister "GeoGuides" series. Projects developed in partnership with UNESCO, World Heritage and Global Geoparks Networks, IUGS and IGU, as well as with the 'Earth Science Matters' Foundation will be considered for publication. This series aims to provide a place for in-depth presentations of developmental and management issues related to Geoheritage and Geotourism in existing and potential Geoparks. Individually authored monographs as well as edited volumes and conference proceedings are welcome; and this book series is considered to be complementary to the Springer-Journal "Geoheritage".

Young Ng · Wolfgang Eder
Editors

The Redstone Park of China

Mount Danxiashan

Springer

Editors
Young Ng
Geotourism Australia
Australian Geoscience Council
Sydney, NSW, Australia

Wolfgang Eder
UNESCO's Division of Earth Sciences
Geoscience Centre, University of Göttingen
Göttingen, Germany

ISSN 2363-765X ISSN 2363-7668 (electronic)
Geoheritage, Geoparks and Geotourism
ISBN 978-3-031-95602-7 ISBN 978-3-031-95603-4 (eBook)
https://doi.org/10.1007/978-3-031-95603-4

This Springer imprint is published by the registered company Springer Nature Switzerland AG
The registered company address is: Gewerbestrasse 11, 6330 Cham, Switzerland

Foreword by Prof. Roland Oberhänsli

Why is Mount Danxiashan, like all other geoparks, of interest to a scientific union such as the International Union of Geological Sciences (IUGS)?

The primary role and mission of the International Union of Geological Sciences (IUGS) are to support broad-based scientific studies relevant to the entire Earth system. The results of these and other studies are applied to preserving Earth's natural environment, using all natural resources wisely, and improving the prosperity of nations and the quality of human life. Founded in 1961, IUGS is one of the world's largest scientific organizations. It is an impartial, not-for-profit scientific union addressing global issues that contribute to the advancement of geological research. The IUGS encourages international cooperation and participation in Earth sciences in relation to human welfare and aims to inform, advise, and influence decision-makers throughout the world. As a member of the International Council for Science (ICS), IUGS has over 100 member countries representing over a million geoscientists. It operates through topic-specific Commissions, Task Groups, and Initiatives, as well as Joint Programmes with other organizations.

Within the UNESCO International Geoscience and Geoparks Programme (IGGP), IUGS parallels the geopark network and is represented by its joint scientific program, the International Geoscience Programme (IGCP). IUGS serves the geopark network by providing experts for geoscience reviews.

Although IUGS's interests include setting international standards on stratigraphy, rock names, geologic GIS maps, and isotopes, the potential of using the natural beauty of geologic phenomena, along with cultural and technological heritage for socio-economic purposes to maintain livelihoods in rural mountainous areas of UNESCO Global Geoparks, aligns perfectly with IUGS's mission. This is especially true as IUGS runs a project called 'Resourcing Future Generations'. Its focus is on the socio-economic effects on resources and questions of demand and supply in a world with a burgeoning population. As large parts of the world in emerging nations strive for better living conditions, the consumption of energy and resources will increase. Population-driven demands will generally increase, and alternative energy to cope with the climate threat requires large grids, significantly multiplying metal needs. Therefore, any goals or visions for the future that ignore mining or assume it will phase out will be seriously compromised. Our societies must understand and cope with the fact that evolution, well-being, and culture all depend on the planet's geology.

Where can geoscience and geoparks meet in this endeavour? It aligns with IUGS's proposed conceptual outline of geoscience professionalism. Decision-makers and citizens need to understand that reserve options must be compared against mineral values. Competing interests for land use, such as forestry, conservation, tourism, defence, mineral exploration, and mining, present a challenge. Sensibilities for our consumption and needs in tomorrow's high-tech world can best be gathered in the beauty of a geopark. It is here that the understanding and appreciation of geologic processes, the impact of local geology on soils, water, and vegetation, as well as regional cultural evolution, can be gathered and hopefully understood without academic lectures. A geopark can convey this knowledge to interested citizens and our children.

The overwhelming beauty of the landscape underscores the importance of geology to our needs and human development. Why else would Maba Man, representing a transitional state from Homo Erectus to early Homo Sapiens in the Middle Palaeolithic period, have chosen this landscape as his dwelling? The water and soil in a spectacular landscape with cave dwellings, exceptional evergreen forests with endemic plants, and game provided the basis for a good life. Similarly, the 'Nian Yu Zhuan' culture and rock carvings from the Tang and Sung Dynasties document that the specific geology of the Mount Danxiashan area presented favourable living conditions for humans.

In my personal view, a geopark has the obligation to utilize its treasures not only to attract tourism but also to show visitors that this is our only liveable planet. Geology is important because 'everything we do not farm, or hunt comes from the Earth'.

Prof. Roland Oberhänsli
Past President of the International Union
of Geological Sciences (IUGS)
Paris, France

Foreword by Prof. Piotr Migoń

It is my great pleasure to write the Foreword for this book, which introduces to the world one of the most magnificent landscapes of China—the scenery of Danxiashan in Guangdong Province.

I had the privilege of visiting Danxiashan twice, thanks to the kind invitation of the late Prof. Peng Hua, a geomorphologist from Sun Yat-sen University and a great promoter of danxia geomorphology. Although both visits took place more than a decade ago, the fond memories of Mount Danxiashan remain vivid, and its impressive, often bizarre landforms are solidly imprinted in my geomorphological mind. I often return to my collection of photographs taken among its red hills to illustrate lectures and conference presentations focused on erosional landscapes in sandstones and conglomerates, of which Mount Danxiashan is such a splendid example. In this way, I also make a modest contribution to popularizing this incredible area.

For a geomorphologist specializing in sandstone terrains, Mount Danxiashan represents an exceptional research opportunity. This opinion is based on the astounding diversity of landforms found in the area. Isolated plateaus and mesas, rock towers and pinnacles, massive domes, deep shaded slots, natural rock bridges, caves and large overhangs, block fields, fluted and hollowed rock faces—all combine into an assemblage of landforms hardly paralleled elsewhere. No wonder Mount Danxiashan has become a reference locality for scholars attempting to decipher patterns of landform evolution in sandstone-dominated terrains. The undoubted scientific value of Mount Danxiashan is enhanced by its aesthetic value, including vivid colour contrasts between red rocks, evergreen subtropical forests, and blue lake and river waters, well-designed tourist paths, panoramic viewpoints, and mysterious hidden corners.

This book includes reviews of recent progress in the scientific understanding of Mount Danxiashan, encompassing not only geology and geomorphology but also ecology and rich cultural heritage. It is heartening to see that research in Mount Danxiashan is ongoing and that our knowledge is expanding. The more we learn about natural areas, the stronger our commitment to their protection and sustainable management. Mount Danxiashan is justifiably a very popular tourist destination. Managing large tourist flows in a fragile environment is invariably a challenge, and Mount Danxiashan is no exception. Solid research should always underpin management plans, and this is clearly the case in Mount Danxiashan, as presented in several chapters within this book. I wish the managers of the Geopark continued success in finding and implementing the right solutions.

Lastly, my visits to Danxiashan were more than just seeing amazing geomorphology. Equally important was the opportunity to meet great people—local staff, park guides, conference organizers and assistants, as well as academic colleagues from China and abroad who, like me, came to see the incredible red rock scenery and share their expertise from similar and not-so-similar landscapes. Some are sadly no longer with us; I have lost contact with others, but I am glad to stay in touch with many, including the editors of this book, who should be congratulated for their efforts. I remember them all, as they were an inseparable part of my Mount Danxiashan experience.

Piotr Migoń
Professor of Geography
University of Wrocław
Wrocław, Poland

Foreword by Edward Yau

There is an old Chinese saying that the greatness of a mountain is not measured by its sheer height but by the immortals who dwell within. Mountains are not merely rock formations but magical creations by Mother Nature, shaped and weathered through time immemorial. They display the spirit and beauty of eternity, often making us humans feel very humble before them.

Mount Danxiashan is no exception. The flame-red sandstone cluster of hills dots the picturesque mountain range in southern China like the paintbrush strokes of autumn maples against the backdrop of a setting sun. Patronized by millions of visitors over the years and centuries, Danxiashan continues to invite and captivate. Among its admirers are renowned geographers and geologists, seasoned travellers, painters, poets, photographers, and those simply seeking a glimpse of the mountains' charm and mystery. There are also worshippers who seek blessings from the hidden immortals.

In recent years, Danxiashan has become more accessible to both local and international visitors with the opening of a high-speed rail link, placing it within a 2–3 h train ride from Guangzhou or even Hong Kong. Despite the influx of visitors, Danxiashan remains a fairyland, a retreat from the hustle and bustle of the highly urbanized province of Guangdong, one of China's economic and manufacturing powerhouses.

Danxiashan is an extraordinary geological wonder beyond its shape and form. As we learn more about how it was formed and crafted through the forces of nature over hundreds of millions of years, we also discover its role as a nurturing home for diverse ecology, and its rich history, culture, religion, and the livelihoods of many generations of people associated with it.

I can attest that the experience of visiting Danxiashan has improved significantly over the last decade. The good practices of geotourism by the park authority have demonstrated how enhanced management can bring the geological attraction's lustre into fuller display, allowing every visitor to take home more than they expected after a single or repeated visit. All these changes merit a deeper understanding and appreciation of the transformation.

This book, co-edited by Dr. Young Ng and Dr. Wolfgang Eder, along with contributions from eminent authors, is a comprehensive collection of research and analyses. It provides a 360-degree review of Danxiashan's geological and geomorphological origin and evolution, as well as its human and economic geographical development from a geotourism perspective. The book aims not only to reach the level of knowledge that appeals to researchers, specialists, and students but also to engage a broader audience. This includes travelers and visitors, nature lovers, park managers, policymakers, tourism and hospitality operators, local inhabitants, and all who are interested in understanding what a Global Geopark like Danxiashan means to them and how they can contribute to and benefit from making the geopark a greater asset to both the environment and the community, in addition to appreciating its scenic beauty.

I am grateful to know Dr. Ng and Dr. Wolfgang, and I hold great admiration and respect for their work in promoting geoparks and geotourism around the world. I am indebted to Dr. Ng, who was one of the pioneers in bringing Hong Kong into the global geopark network. Through his passion, advocacy, and tireless efforts, Hong Kong's coastal volcanic formation attained the status of national and global geopark in 2008 and 2009, respectively, during my tenure as Hong Kong's Secretary for Environment. It was also through the process of securing UNESCO's naming and Dr. Ng's introduction that I came to know Dr. Wolfgang and appreciate his leadership in the geopark movement. Dr. Ng and Dr. Wolfgang are indeed the immortals who reveal the hidden geo treasures, displaying their new charm before us.

I recommend you read this book and even take it with you when you visit Mount Danxiashan in the future. I certainly will.

Hong Kong, China

Edward Yau
GBS, JP Hong Kong's Secretary for
Environment (2007–2012) and Secretary for
Commerce and Economic Development
(2017–2022)

Foreword by Prof. David Newsome

When I visited Mount Danxiashan Geopark in 2008, I knew I was witnessing an extraordinary landscape—one that holds cultural and aesthetic significance stretching back through time. The international importance of Mount Danxiashan became particularly evident in the Asia Pacific region some years ago, when a 'sister park' agreement was signed between the Department of Parks and Wildlife Service in Western Australia and Guangdong Mount Danxiashan National Park Management. This agreement celebrates two World Heritage properties of great geological significance and tourism value. Purnululu National Park is Australia's 'danxia' landscape, typified by its orange/red sandstone towers commonly referred to as 'beehive' structures. China, however, is the true home of danxia landscapes, with the origin being Mount Danxiashan itself.

From a personal perspective, there are two themes of particular significance that I wish to reflect upon. The first is the importance of learning and the educational role that geoparks must play. I often use the analogy of meeting someone for the first time. Initially, you don't know anything about them—where they come from or how they might inspire those around them. However, as you get to know this person, it opens the opportunity for a deeper relationship and a more meaningful experience. Learning about geology and landscapes is similar; understanding landforms and rock outcrops can answer questions of origin, appearance, and, always intriguing, questions of time and age.

The second theme is the need for careful management. We live in a time when being outdoors and visiting natural areas is more important than ever. This is evident from the increasing visitor numbers to various protected areas and geoparks worldwide. People instinctively know that nature is good for them, and there is growing scientific evidence that being in and learning about the natural world is essential for our mental and physical well-being. However, over-tourism has emerged as a threat to the integrity of many protected areas, potentially spoiling the very experience people seek when they view a landscape or spend time in a natural setting. Hence, the vital importance of visitor management. We must carefully consider how we market geoparks as leisure destinations, control visitor numbers, and exercise mindfulness regarding what is and isn't allowed in places like Mount Danxiashan, the Redstone Park of China.

Mount Danxiashan is also a World Heritage site for all to enjoy, and this book encapsulates the knowledge that enhances that enjoyment. I commend this book as an important contribution to geotourism literature and eagerly look forward to my next visit.

David Newsome
Murdoch University
Perth, Australia

Preface

Over the past three decades, environmental geoscience programs and governmental management institutions worldwide have significantly evolved. These changes reflect the need and responsibility to provide fundamental geoscience information to the public and political decision-makers regarding nature and society. This encompasses sustainable resource development, nature assessment and protection, natural hazards monitoring and prediction, global climate change, policy generation, public safety, education, and tourism. Since 1994, the topic 'Geoscience in a Changing Society' has been raised by many national geological survey offices. The term 'social geology' was first introduced by Oiti Berbert (Brazil) at the 30th International Geological Congress in Beijing in 1996, forming the foundation of UNESCO's 'Global Geoparks Programme'. The general topic of 'value of nature' has since evolved into a crucial question of whether the costs of projects are justified by the returns on the investment.

In addition to its strong ties with culture and education, geo has evolved into a novel marketing product. Colleagues from the Chinese Academy of Geological Sciences, including Xun Zhao and Anze Chen, along with Prof. Hua Peng from Zhongshan University, Guangzhou, were among the first to recognize the educational and economic opportunities of the outstanding geomorphological landscape of Mount Danxiashan. They played their cards well, modeling numerous national and provincial geoparks in China with excellence.

The geopark concept is intrinsically linked to geoheritage conservation and sustainable development. The movement began in Europe in the 1990s by a few 'geo-enthusiasts'—geologists, paleontologists, geographers, and environmental planners—who recognized that the conservation and development of significant geological and geomorphic sites could serve as an educational tool for understanding Earth, planets, and life. It also presented a promotional tool for regional and local sustainable economic development through 'geotourism'.

Since 1994, UNESCO's former Division of Earth Sciences, under the directorship of Wolfgang Eder (co-editor of this book), explored the possibilities of creating a global network of selected geoheritage-related territories under UNESCO's umbrella to promote the understanding of our Earth. At the 30th International Geological Congress (Beijing 1996), a workshop on 'Geological Heritage and Geosites/Geoparks' was organized by Mechtild Rössler of UNESCO's World Heritage Centre, Paul Dingwall of New Zealand, Xun Zhao of China (also Secretary General of the 30th IGC), and Wolfgang Eder. This initiative was supported and driven by China's former Ministry of Land and Resources, along with the Academy of Geological Sciences represented by Xun Zhao, Anze Chen, and Jianjun Jiang.

In 2004, the Global Geoparks Network (GGN) was established in Beijing, with support from European and Chinese geoparks, and Mount Danxiashan was one of its founding members. Wolfgang Eder had the privilege of visiting multiple times since 2003. The spectacular danxia landscape was formed on continental red terrigenous sedimentary beds influenced by tectonic uplifting, followed by weathering and erosional processes. The breathtaking wall cliffs, canyons, beautiful forests, and lakes, together with rich cultural heritage, earned it World Heritage status as 'China Danxia' along with five similar landscapes in China.

The first idea of compiling a bilingual (English/Chinese) book on 'Mount Danxiashan' was initiated during a river cruise organized in 2007 by the City of Shaoguan. Professor Hua Peng, after performing a few songs, approached the mayor of Shaoguan and other representatives of

Guangdong Province. He convinced them to continue their invaluable geo-related support for Mount Danxiashan by publishing a book to reach the international community. The original English manuscript by Prof. Hua Peng took a long time to materialize, even with assistance from Peter Bobrowsky (Canada), Chris Wood (UK), Paul Dingwall (New Zealand), and Tim Kusky (USA). However, with Peng Hua's passing and rapid changes in global geopark development, the incomplete manuscript lacked updated data and references. It did not meet the strict requirements for publication and global distribution. Thus, the idea of publishing a book based on Hua Peng's original manuscript was set aside.

To continue introducing Mount Danxiashan to the world, Young Ng initiated the idea in 2021 of publishing a book based on the latest research and works of current frontline researchers and managers of Danxiashan. Each invited contributor wrote a chapter in their area of expertise. These contributors included Mr. Yuan Zheng, ex-Director of the Chinese Geoparks Network (Chapter 'Chinese Geopark Development'), Profs. Liuqin Chen, Fusheng Guo and Fujun Liu (Chapter 'Geology'), Prof. Zhang Ke (Chapter 'Geomorphology'), Profs. Wenbo Liao and Ting Zhou (Chapter 'Ecology'), Mr. Rongfeng Hou (Chapter 'History and Culture'), Ms. Fang Chen and Mr. Yuqiang Li (Chapter 'Management and Sustainable Development'), Prof. Luobin Yan (Chapter 'Education and Research'), and Drs. Yao Meng and Xin Li (Chapter 'Geoscience Popularization'). Through their contributions, all aspects of Mount Danxiashan were systematically covered.

This book provides an overview of Mount Danxiashan, from its original objectives of establishing the geopark, its geological, geomorphological, and ecological uniqueness, to its history, culture, management, research, and science popularization. It serves as an excellent reference for those interested in geopark and national park development in China, particularly for those keen to explore how China sustainably uses its geological heritage to protect the environment while improving the livelihood of local communities. The book also caters to tourists interested in visiting a typical Chinese geopark, appreciating the fabulous red sandstone landscape, and experiencing the integration of Chinese culture with nature.

This book is dedicated to the memory of the late Prof. Hua Peng, whose profound passion and decades of tireless research in danxia landscapes have greatly advanced our understanding of these unique geological formations. We also honor the late Prof. Luobin Yan, who, with great sorrow, passed away during the preparation of his chapter. His invaluable contributions and unwavering commitment are deeply missed.

We extend our heartfelt gratitude to all the contributors for their time and effort in meticulously preparing each chapter. Special thanks go to our colleagues at the Danxiashan Research Institute, the Guangdong Administration of Coal Geology, and the Danxiashan Administrative Committee of Shaoguan City for their tremendous support, without which this publication would not have been possible.

Lastly, and most importantly, we extend our sincere thanks to our esteemed colleagues and friends from around the world who graciously contributed their forewords to this book. Their visits to Mount Danxiashan and their unwavering love for its remarkable landscape, coupled with their steadfast support for our work, have been a source of immense encouragement and inspiration.

We are deeply honored by the contributions of Prof. Roland Oberhansli, Past President of the International Union of Geological Sciences (IUGS), Prof. Piotr Migoń from the University of Wroclaw, Poland, Mr. Edward Yau, former Secretary for Commerce and Economic Development of the Hong Kong SAR Government, and Prof. David Newsome from Murdoch University, Western Australia.

Their insightful perspectives and enduring support not only enrich this book but also reflect their shared commitment to celebrating and preserving the extraordinary heritage of Mount Danxiashan. For this, we are profoundly grateful.

Sydney, Australia — Young Ng, Ph.D.
Göttingen, Germany — Wolfgang Eder, Ph.D.

Contents

Background

Young Ng and Wolfgang Eder

Abstract

Located in the northeast of Shaoguan City in Guangdong Province, Danxiashan UNESCO Global Geopark, also known as the Redstone Park of China, spans a total area of 290 km^2. About half of this area is densely covered by the iconic 'Danxia landform'. The highest peak in the park is Bazhai, reaching an elevation of 619 m. Mount Danxiashan had been designated as a national scenic area in 1988 and a national Geological and Geomorphological Nature Reserve in 1995 (State Council 1988, 1995). It became a UNESCO Global Geopark in 2004 and a World Heritage Site in 2010 (UNESCO 2024, 2025).

1 Background

Located in the northeast of Shaoguan City in Guangdong Province, Danxiashan UNESCO Global Geopark, also known as the Redstone Park of China, spans a total area of 290 km^2. About half of this area is densely covered by the iconic 'Danxia landform'. The highest peak in the park is Bazhai, reaching an elevation of 619 m. Mount Danxiashan had been designated as a national scenic area in 1988 and a national Geological and Geomorphological Nature Reserve in 1995 (State Council 1988, 1995). It became a UNESCO Global Geopark in 2004 and a World Heritage Site in 2010 (UNESCO 2024, 2025) (Figs. 1 and 2).

Characterized by multi-layered red sedimentary rocks of sandstone and conglomerate, the area was formed in a basin by fluvial deposition between 140 and 65 million years ago. The warm and wet climate enhanced oxidation in the rocks, turning the colour of their sediments to red. These rocks were later uplifted, undercut by water, and accelerated by rockfall, resulting in the present Danxia landscape. This region is geologically significant for studying the evolution of Earth's crust, featuring rock outcrops from various geological periods, including the Paleozoic, Mesozoic, and Neozoic eras. The Cretaceous sedimentary strata that form the current landscape are particularly important for studying dinosaur extinction, mammal evolution, and regional stratigraphical correlation (Fig. 3).

Mount Danxiashan has undergone numerous geological processes throughout its history, leading to its distinctive formation. It is currently in the mature stage of landform development, with deposition, uplifting, weathering, and erosion occurring simultaneously (Huang 1996, 2004; Peng et al. 2004, Peng and Pan 2013a, b). This dynamic process characterizes the diversified shapes and forms of this unique type of landscape, known in China as the 'Danxia landscape'. As the most representative and naming site of the Danxia landscape, Mount Danxiashan has been studied for nearly a century by Chinese geoscientists, resulting in over 300 academic papers. It has become a global study centre of red beds, especially in warm and humid regions. Its special geology and geomorphology have created a unique ecological environment for diverse flora and fauna. It is a research hub for studying the evolution of the 'Bio-Island Effect', complex differentiations of vegetation, and ecosystems in highly isolated hills and mesas environments.

The unique landscape and humid monsoon climate support evergreen broadleaf forests and various endemic, endangered, and newly discovered species such as Danxia Viola, Danxia Orchid, Danxia Firmiana, and Danxia Chiritopsis. Mount Danxiashan also boasts significant natural aesthetic value due to its rivers (Jinjiang and Zhenjiang), red peaks (Yangyuanshi

Y. Ng (✉)
Geotourism Australia, Australian Geoscience Council, Sydney, Australia
e-mail: oz.geotourism@gmail.com

W. Eder
Geoscience Centre, University of Göttingen, Göttingen, Germany

Y. Ng and W. Eder (eds.), *The Redstone Park of China*, Geoheritage, Geoparks and Geotourism,
https://doi.org/10.1007/978-3-031-95603-4_1

Fig. 1 Sunrise at the iconic Zengmaofeng (Monk Hat) Peak of Mount Danxiashan

and Zengmaofeng), cliffs (Chahufeng and Guanyinshi), caves (Longlinyan), canyons, and valleys (Bazhai).

2 Landform or Landscape?

Most Chinese scholars use the term 'Danxia landscape' (丹霞地貌) to describe the overall terrain, rather than individual geological and geomorphological features within the Mount Danxiashan area. It also includes the vegetation and cultural influences that have played an important role in shaping the landscape. Therefore, the term 'Danxia landscape' is more appropriate in English than 'Danxia landform' when describing the overall landscape of an area. Adopting this correct English terminology in future references will help to avoid any misinterpretations and will help popularizing the concept among researchers and the public within and outside China.

The term 'Danxia landscape' refers to a unique terrain that bears a resemblance to karst landscape. However, in China, the terms 'landform' (地形) and 'landscape' (地貌) in natural environments are often misapplied, largely due to cultural misperception of geological and geomorphological characteristics by blending the concepts of landform and landscape when describing terrains. To clarify the distinction between these terms:

***Landform*:** A landform is a natural feature of the Earth's surface. Examples include a mountain, a waterfall, a cliff, or a sand dune. These features are specific, individual elements that make up the Earth's topography.

Landscape: A landscape encompasses an area with various natural features or landforms. Examples include a river valley, an alluvial plain, a coastline, or a desert. For instance, a desert landscape can contain numerous landforms such as dunes, ergs, mesas, buttes, wadis, oases, salt flats, and playas. Thus, landforms are subsets of landscapes, representing individual features within a broader geographical context.

When describing a special area with a cluster of stunning cliffs, waterfalls, caves, and river meanders like Mount Danxiashan, it is more accurate to use the term 'Danxia landscape" rather than 'Danxia landform', unless referring to a specific landform or the various landforms within the area.

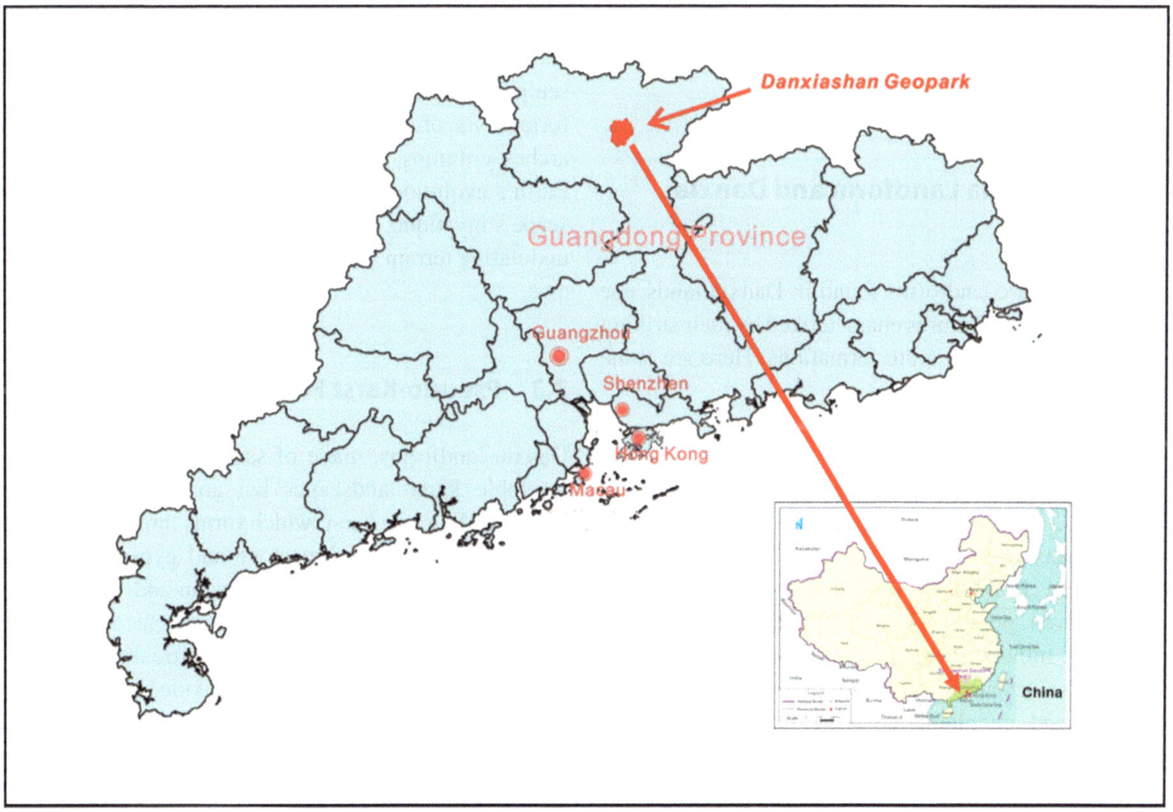

Fig. 2 Location of Mount Danxiashan and Danxiashan UNESCO Global Geopark, Guangdong Province, China

Fig. 3 The multi-layered sandstone and conglomerate cliff is a distinctive feature of Mount Danxiashan

In addition to describing geological and geomorphological features, the term 'landscape' can also be used to characterize the ecological and cultural environments. Hence, we use 'ecological landscape' to describe the vegetation and habitats, and 'cultural landscape' to depict the historical and cultural characteristics of an area.

While landforms refer to the physical shapes or landforms on the Earth's surface created by natural processes, landscapes encompass both these landforms and the cultural, ecological, and historical significance attached to them. Danxia landforms are characterized by red-coloured sandstones and conglomerates, which have been eroded over time to form steep cliffs and unique rock formations. The Danxia landscape, however, includes the broader environmental context, including flora, fauna, human history, and cultural narratives that have evolved around these distinctive landforms.

Therefore, the term 'Danxia landscape' is more appropriate to be used to describe the overall terrain of Mount Danxiashan (Fig. 4).

3 Defining Danxia Landform and Danxia Landscape

Danxia landforms are landforms found in Danxia landscape with unique geological features characterized by their striking red sandstone and conglomerate formations. Here are some of their distinctive characteristics.

3.1 Colourful Rock Formations

Danxia landforms are celebrated for their strikingly vibrant colours, ranging from deep reds and oranges to bright yellows and even purples. These captivating hues are a direct result of mineral deposits and the oxidation of iron within the rock formations. Over millions of years, layers of sandstone and conglomerate have been subjected to various geological processes, leading to the oxidation that produces these vivid colours. This palette of warm tones not only enhances the visual appeal of Danxia landscape but also provides valuable insights into the Earth's geological history.

The unique coloration of Danxia landforms creates a stunning contrast against the surrounding environment, making these landscapes a visual feast for visitors and an important subject of study for geologists and researchers. The dramatic interplay of light and shadow on the brightly coloured cliffs and valleys further accentuates their beauty, especially during sunrise and sunset. These natural wonders are not just a feast for the eyes but also a testament to the Earth's dynamic geological processes, offering a glimpse into the forces that have shaped our planet (Fig. 5).

3.2 Steep Cliffs and Towers

Danxia landforms are distinguished by their remarkable geological features, which often include flat-topped summits, steep cliffs, towering rock formations, and isolated peaks. These elements come together to create a dramatic and rugged terrain that leaves a lasting impression on all who witness it. The flat-topped summits, reminiscent of ancient fortresses, rise prominently above the landscape, offering breathtaking panoramic views. These elevated plateaus provide a stark contrast to the precipitous cliffs that drop sharply to the valleys below, emphasizing the vertical relief and grandeur of the terrain (Qi et al. 2015; Zhu and Peng 2009).

The towering rock formations, with their intricate shapes and vibrant colours, stand as sentinels of the geological past, sculpted by the forces of nature over millions of years. These formations often include striking features such as natural arches, columns, and spires, each telling a unique story of the Earth's evolutionary history. Isolated peaks add to the landscape's mystique, appearing as solitary guardians amidst the undulating terrain, further enhancing the sense of wonder and awe.

3.3 Pseudo-Karst Features

Danxia landforms, made of sandstones and conglomerates, resemble karst landscapes but are classified as 'pseudo-karst'. Unlike true karst, which forms through the dissolution of soluble rocks like limestone and gypsum, Danxia landforms result from differential erosion and weathering of red sandstone. This process gives rise to spectacular cliffs, gorges, and columnar rock formations, often streaked with vivid hues of red and orange due to the iron oxide content in the rocks. These unique landscapes are named after Mount Danxiashan in China and can be found in various regions across the world, including China, the United States, and Australia. The vibrant layers and dramatic shapes make Danxia landforms a visually stunning and geologically intriguing feature of the natural world.

3.4 Unique Caves

Unlike true karst terrains with deep, interconnected cave systems, Danxia landforms have shallow, isolated caves formed by weathering and erosion. These caves, often small and scattered, are shaped by the gradual breakdown of the red sandstone and conglomerates that make up Danxia landscape. This weathering is primarily caused by the action of wind, water, and temperature fluctuations, leading to the creation of various cave types, including honeycomb caves and grottoes. These formations are often adorned with unique features such as natural arches, rock pillars, and stalactite-like structures, making them fascinating sites for geological study and exploration. Despite their modest size, these caves contribute to the overall beauty and complexity of Danxia landforms, offering intriguing glimpses into the geological processes at work (Fig. 6).

3.5 Geological History

Danxia landforms were shaped by tectonic uplift and subsequent erosion over millions of years, exposing layered rock

Fig. 4 Ariel view of Mount Danxiashan (*Source* Baidu Images https://image.baidu.com/)

Fig. 5 Jinjiang River cut across Mount Danxiashan to unveil the steep cliff of Jinshiyan Rock

formations. The process began during the Cretaceous Period when vast quantities of red sandstone and conglomerates were deposited in sedimentary basins. As tectonic forces caused the Earth's crust to uplift, these sedimentary layers were pushed upward, forming the basis of the different Danxia landscapes we see today. Following the tectonic uplift, the exposed rocks were subjected to continuous weathering and erosion. This erosion was driven by a combination of natural forces, including wind, rain, and temperature fluctuations, which gradually sculpted the landscape into its current form. Over time, the softer rock layers eroded more quickly than the harder ones, leading to the formation of cliffs, gorges, and isolated rock pillars.

The geological history of Danxia landforms is also marked by the presence of distinct stratigraphic layers, each representing different periods of sediment deposition. These layers provide valuable insights into the ancient environments and climatic conditions that existed millions of years ago. The vibrant red and orange hues of the rocks are a result of iron oxide within the sandstone, which has oxidized over time, adding to the visual appeal of the landscape. In summary, the geological history of Danxia landforms is a testament to the dynamic forces of nature. The interplay between tectonic uplift and erosional processes has created a unique and visually stunning landscape that continues to captivate geologists and visitors alike.

Fig. 6 Steep cliff with grooves created by water erosion in Danxia landscape (Picture: Hou R.F.)

3.6 Ecological Significance

Danxia landforms, particularly in regions with warm and humid climates like Mount Danxiashan, create unique habitats that support a diverse array of flora and fauna. These conditions have led to the discovery of rare and endemic species that hold significant ecological importance. The interplay between the distinctive geological features and the region's climate fosters a rich biodiversity, making these landscapes critical for conservation efforts and ecological studies.

3.7 Cultural Significance

Many Danxia landscapes both in China and outside China, hold profound cultural and historical significance, often deeply embedded in the local legends and traditions of the regions they inhabit. These unique formations have not only shaped the physical landscape but have also influenced the cultural narratives and heritage of local communities. For centuries, Danxia landscapes have been revered in folklore, with their striking natural beauty inspiring countless myths and legends that have been passed down through generations. Temples, shrines, and ancient inscriptions found within these landforms stand as testaments to their sacred status and historical importance. Festivals and rituals celebrating the natural splendour of Danxia landscape continue to be integral parts of local traditions, reflecting the enduring connection between the people and their environment. This rich tapestry of cultural and historical elements highlights the multifaceted significance of Danxia landscape, making them invaluable not only for their geological features but also for their contributions to cultural heritage and identity.

3.8 Defining Danxia Landscape

A Danxia landscape is therefore can be defined as a unique and comprehensive landform group that developed in the South China Plate and the humid subtropical climate zone of Central and South China. It is characterized by its formation within the red continental sandy conglomerate strata of the Middle and Upper Cretaceous series, shaped predominantly by processes such as water erosion, dissolution, and gravitational forces. This geological evolution gives rise to striking landform features, including red precipices, mesas, stone walls, stone peaks, pinnacles, canyons, ravines, arches, and caves. Exemplified by Mount Danxiashan in Guangdong, China, it represents a distinctive category of red bed landforms (Chen 2016; Peng and Pan 2013a, b).

From a broader geographical perspective, when describing an area that includes various Danxia landforms, along with the unique ecosystem and culture they produce, it can be regarded as a landscape like other types such as karst, coastal, and desert landscapes. The authors describe the geomorphological features of Mount Danxiashan as an area of 'Danxia landscape' unless referring its specific landforms throughout the book (Fig. 7).

4 Typical Danxia Landscape: Mount Danxiashan

Mount Danxiashan, situated in the heart of Guangdong Province, stands as a premier representation of Danxia landscapes (Chen 2005, 2016; Peng et al. 2004; Peng and Pan 2013a, b; Qi et al. 2015). This distinction is primarily attributed to the site's impeccably preserved geological

Fig. 7 One of the pinnacles formed by collapse of top layers of rocks due to erosion along joints of the damxia landscape

formations, which exhibit a striking array of red sandstone peaks, cliffs, and valleys that captivate geologists and visitors alike. Additionally, the region boasts an extraordinary diversity of flora and fauna, making it a haven for biodiversity. The significance of Mount Danxiashan extends beyond its natural features, as it is deeply intertwined with the cultural and historical narratives of the region. Over the centuries, these natural wonders have inspired a rich tapestry of local traditions, legends, and historical events, further enriching the area's cultural heritage. The harmonious blend of stunning natural beauty and profound cultural heritage at Mount Danxiashan offers a unique and comprehensive model for understanding the intricate relationship between geological formations and cultural landscapes. This integration serves as a testament to the enduring legacy of nature's influence on human history and culture (Fig. 8).

5 UNESCO Branding: Enhancing Tourist Attractions

Mount Danxiashan's designation as a UNESCO Global Geopark and World Heritage Site under China Danxia serves as a cornerstone in augmenting its attractiveness to tourists from all corners of the globe (UNESCO 2024, 2025). This prestigious UNESCO brand acts as a beacon of international recognition, highlighting the profound geological and cultural importance of the site. Such recognition not only piques the interest of travellers seeking unique and enriching experiences but also attracts scholars and researchers eager to delve into the area's rich natural history (Weng and Peng 2014). The endorsement by UNESCO significantly enhances the site's reputation, positioning it as a must-visit destination for both leisure and academic pursuits. Moreover, this global acknowledgment underlines the necessity of adopting sustainable tourism practices. By promoting responsible tourism, the UNESCO branding ensures that the delicate ecosystems and invaluable geological features of Mount Danxiashan are meticulously preserved, allowing future generations to explore and appreciate its exceptional beauty and scientific significance.

6 Reasons for Publishing This Book

The publication of this book is driven by a deep-seated desire to shed light on the remarkable Danxia landscape, whose unique natural formations stand as a testament to Earth's dynamic geological history. With its outstanding geology and geomorphology, the Danxia landscape offers invaluable insights into the planet's evolutionary processes, particularly during the late Cretaceous period. Despite being recognized as a World Heritage Site, the true significance of Danxia landscape remains underappreciated, especially beyond the borders of China.

This book aims to bridge that knowledge gap, bringing the scientific importance and breathtaking beauty of Danxia landscape to the forefront of global academic and public awareness. By compiling comprehensive research and vivid descriptions, we hope to highlight the crucial role these landscapes play in understanding Earth's past. This endeavour not only honours the natural heritage of the Danxia landscape but also seeks to inspire further exploration and appreciation of these geological wonders worldwide.

Fig. 8 The main entrance to the Danxiashan Global Geopark

This book offers a comprehensive introduction to Mount Danxiashan, China's most renowned red stone park and one of the world's first UNESCO Global Geoparks. It is enriched by the insights and knowledge shared by experienced researchers, educators, and park managers who have dedicated their careers to studying Mount Danxiashan. The book comprises 9 chapters, beginning with an introduction and concluding with a final epilogue.

In this Chapter: The introductory chapter outlines the background of Mount Danxiashan and the rationale behind the creation of this book.

Chapter 2*:* Authored by Mr. Yuan Zheng, the former director of the Chinese Geoparks Network (CGN), this chapter provides an overview of geopark development in China, with a particular focus on Mount Danxiashan. Mr. Zheng, who has over two decades of experience in coordinating and assessing geoparks for both the CGN and the Global Geoparks Network (GGN), brings invaluable expertise to this chapter.

Chapter 3*:* Professor Liuqin Chen delves into the geology of Mount Danxiashan, offering readers a fundamental understanding of the geological processes that have shaped the Danxia landscape.

Chapter 4*:* Professor Ke Zhang from Sun Yat-sen University explores the landforms contributing to the stunning appearance of Mount Danxiashan, detailing the erosion and weathering processes that have sculpted the landscape.

Chapter 5*:* This chapter covers the ecology of Mount Danxiashan, authored by Professors Wenbo Liao and Ting Zhou. Along with their research team from Sun Yat-sen University, Professor Liao has decades of experience working with the region's flora. Their ongoing efforts have led to the discovery of new plant species, highlighting the area's ecological significance.

Chapter 6*:* Mr Rongfeng Hou, a cultural expert and deputy director of Danxiashan UNESCO Global Geopark, writes about the rich history and culture of Mount Danxiashan. This chapter emphasizes the intertwining of natural beauty and cultural heritage in the region.

Chapter 7*:* Fang Chen and Yuqiang Li of Danxiashan UNESCO Global Geopark, shares invaluable insights on the management and sustainable development of the geopark. She details her persistent efforts to elevate Mount Danxiashan to one of China's premier geoparks.

Chapter 8*:* The late Professor Luobin Yan of Southwest University of Szechuan presents the role of research and education within the geopark. His chapter summarizes the significant work done in Mount Danxiashan. Although Professor Yan passed away before completing the chapter, Dr Young Ng, the Chief Editor, has respectfully finished it in his memory, honouring his contributions.

Chapter 9: Dr Yao Meng of China University of Geosciences, along with his colleague, Dr Xin Li, director of the Geoparker Natural Science Popularization Centre discusses the science popularization efforts and educational activities conducted in Mount Danxiashan over the past 15 years. Their chapter underscores the importance of public engagement and education in the ongoing conservation and appreciation of the geopark.

The editors believe that this book comprehensively covers all aspects of Mount Danxiashan, providing a complete view of the geopark. It delves into not only the striking Danxia landscape but also the ecology, culture, research, education, management, and science popularization efforts. These elements collectively make Danxiashan UNESCO Global Geopark unique and special.

References

Chen AZ (2005) Six principles of landform naming. In: The 11th collection of tourism earth science. China Tourism Earth Science and Geoparks Research Committee, Geological Society of China, Beijing

Chen AZ (2016) Discussion on several issues of Danxia landforms. In: Collected papers from the 19th national tourism geology annual meeting and Shaoguan City Tourism Development Strategy Seminar. Shaoguan, China

Huang J (2004) Quantitative measurement of several important issues in the development of Danxia landform. In: Huang J (eds) Danxia landscape in China. Science Press, Beijing, pp 66–74 (2010)

Huang J et al (1996) A preliminary study of Danxia cave geomorphology. In: Theory and practice of geotourism—collection of geotourism papers, vol II, pp 32–45

Peng H, Pan ZX (2013a) A review on the research of China Danxia geomorphology in the past eighty years. In: Sandstone landscape, proceedings of the 3rd international conference on sandstone landscape. Wroclaw University, Poland, pp 133–139

Peng H, Pan ZX (2013b) Study on the weathering pattern and evolution process of Danxia bedding caves. In: Sandstone Landscape, proceedings of the 3rd international conference on sandstone landscape. Wroclaw University, Poland, pp 140–145

Peng H et al (2004) Academic values of Danxiashan Geopark and the significance of the international comparison of Danxia geomorphology. In: Proceedings of the first international conference on geoparks. Geological publishing House, pp 123–130

Qi DL et al (2015) Evolution and geomorphological age of Danxia landforms in Kongtong Mountain. J Mt Sci VII:33–42

State Council, PRC (1988) Notice of the State Council on Announcing the Second Batch of National Key Scenic and Historic Interest Areas. State Council Document 1 August 1988 No. 51. State Council's policy database for historical bulletins, Beijing

State Council, PRC (1995) Notice of the State Council on Publishing the List of National Nature Reserves, State Council Document (1995), No. 113, Beijing

UNESCO (2024) Danxiashan UNESCO Global Geopark. UNESCO Global Geopark Official website. https://www.unesco.org/en/iggp/Danxiashan-unesco-global-geopark. Accessed 18 Nov 2024

UNESCO (2025) China Danxia. UNESCO World Heritage Convention. https://whc.unesco.org/en/list/1335/. Accessed 11 Jan 2025

Weng S, Peng H (2014) Tourism development, rights consciousness and the empowerment of Chinese historical village communities. Tour Geogr 16(5):772–784

Zhu C, Peng H et al (2009) Age and genesis of the Danxia landform on Mount Jianglangshan, Zhejiang Province. J Geogr Sci 19(5):615–630

Chinese Geopark Development

Yuan Zheng

Abstract

Mount Danxiashan earned its global geopark status in 2004, making it one of the first global geoparks established worldwide. Its mission and development journey serve as a model for the evolution of many geoparks in China. The Chinese government has been a steadfast supporter of geoparks, not only for preserving geological heritage but also for advancing geoscience education and fostering sustainable development. These geoparks go beyond geology, encompassing integrated ecological and cultural environments that thrive within their geological context. In China, geoparks play a pivotal role in enhancing local livelihoods, thanks to substantial government backing.

1 Background and Origin

Mount Danxiashan earned its global geopark status in 2004, making it one of the first global geoparks established worldwide. Its mission and development journey serve as a model for the evolution of many geoparks in China. The Chinese government has been a steadfast supporter of geoparks, not only for preserving geological heritage but also for advancing geoscience education and fostering sustainable development. These geoparks go beyond geology, encompassing integrated ecological and cultural environments that thrive within their geological context. In China, geoparks play a pivotal role in enhancing local livelihoods, thanks to substantial government backing (Fig. 1).

Geoparks in China are founded on the principle that 'UNESCO Global Geoparks are single, unified geographical areas where sites and landscapes of international geological significance are managed with a holistic approach to protection, education, and sustainable development' (UNESCO 2025).

The concept of geoparks emerged from a renewed understanding and sustainable use of Earth's resources. In the late 1990s, the idea of geoparks—aimed at protecting geological relics, promoting earth science education, and fostering local economic development—arose almost simultaneously in China and Europe. The practices of developing geoparks in both regions reflect this shared vision.

China's journey in geotourism and geopark development dates to the mid-1980s. In November 1985, the 'First Geological Nature Reserve Zoning and Scientific Investigation Work Conference' was held in Changsha, Hunan Province. This conference addressed the unique Devonian sandstone pillar landforms, emphasizing the protection of natural landscapes while supporting the rapidly growing tourism industry. Scholars proposed the creation of comprehensive national nature reserves (or national geological parks) that showcased geological features, landforms, flora, and fauna. This marked a shift in focus for geologists—from traditional mine prospecting to serving societal and tourism needs—while recognizing geological relics and landforms as non-traditional resources.

Two years later, the Ministry of Geology and Mineral Resources issued the 'Notice on the Establishment of Geological Nature Reserves (Trial Implementation)', which introduced departmental regulations for protecting geological relics. In May 1995, the Ministry issued the 'Regulations on the Protection and Management of Geological Relics' (Minister Order No. 21). Article 8 of this regulation explicitly stated that geological relics of international, national, and regional importance should be protected within designated geological heritage protection areas (MLR 1995; Ji and Pan 1997). This document emphasized that geoparks were

Y. Zheng (✉)
Chinese Geoparks Network, Chinese Academy of Geological Sciences, Beijing, China
e-mail: zhengyuan8819@sina.com

Y. Ng and W. Eder (eds.), *The Redstone Park of China*, Geoheritage, Geoparks and Geotourism,
https://doi.org/10.1007/978-3-031-95603-4_2

Fig. 1 Distribution of global geoparks around the world in 2025 (*Source* http://www.globalgeopark.org/GeoparkMap/geoparks/China/index.htm)

Fig. 2 Logos of UNESCO Global Geopark and National Geopark of China

primarily created to safeguard geological relics and landscapes. While geotourism was encouraged, geoparks were still classified within the broader category of nature reserves (Fig. 2).

On an international scale, the concept of geoparks traces its origins to the 1990s. During this period, the need to preserve and enhance the value of geologically significant areas became a key focus (Eder 1999; Wimbledon et al. 1998, 1999). Geological formations and landscapes—serving as markers of Earth's evolution—were recognized for their essential role in sustainable development. Efforts to protect such key areas, which represent the 4.6-billion-year history of Earth, inspired the concept of geological heritage. This concept incorporated landscape protection, geoscience education, and geotourism, with active participation from local communities and stakeholders. These core values eventually gave rise to the modern concept of geoparks, emphasizing conservation, education, and community engagement.

In 1991, the International Symposium on the Conservation of the Geological Heritage was jointly organized by the International Union of Geosciences (IUGS) and the International Union of Geographers (IUG) in Digne, France. The symposium strongly resonated with the emerging ideas surrounding geological heritage preservation. During the event, participants discussed and adopted the International Declaration on the Rights of the Memory of the Earth, which was also being called the Digne Declaration, introducing the concept of 'sharing our common heritage with the Earth'. (Jones 2008; UNESCO 2024). The declaration emphasized that geological relics are part of humanity's shared

Fig. 3 Dunhuang Yardang National Geopark, Kansu Province, China (*Source* Zhang Y.)

heritage and that their protection is a collective responsibility. Another key outcome of this symposium was the global support for geological relic preservation, following its establishment of the Man and Biosphere Program and the World Heritage Program in the early 1970s. The event was attended by more than 120 geologists from over 30 countries, including China.

With growing international attention, the 30th International Geological Congress, held in Beijing in 1996, became a pivotal milestone in geopark development. Under the theme 'Continental Geology—Resources, Environment, Disaster Reduction, and Development', the congress sought innovative solutions to pressing geoscience issues and aimed to strengthen the societal relevance of geology (Zhang 1997; Zhang and Zhang 2022). The congress specifically included sessions on geological relic preservation and organized more than 80 geotourism routes. During a visit to the Zhoukoudian site in Fangshan, experts deliberated on topics such as geological heritage conservation, community engagement, geoscience tourism, and the potential establishment of a European geopark network.

UNESCO has demonstrated consistent support for geoparks from their inception. Although the 161st meeting of the UNESCO Executive Committee in June 2001 did not formally approve the creation of a geopark program, it endorsed the development of unique geological feature areas or natural parks proposed by member states and promoted their construction. In 2002, UNESCO's Department of Earth Sciences published the Guidelines for the Network of Global Geoparks, which redefined the global geopark concept, clarified the conditions and application procedures for geopark designation, and offered a foundation for entities seeking inclusion in the UNESCO Global Geopark Network. These guidelines provided essential principles for the establishment and management of geoparks (UNESCO 2015). The timely release of these guidelines ultimately facilitated the birth of the Global Geopark Network (Fig. 3).

2 Development History

Reflecting on the development of China's geoparks, their objectives and mission have evolved over time. Various definitions of national geoparks have emerged throughout this journey. The Guidelines for the Construction of China's National Geoparks, published in 2006, define a national geopark as a 'unique natural area of significant geological scientific importance, rare natural attributes, and exceptional aesthetic value'. Such parks are characterized by their geological relic landscapes within a specific area, integrated with other natural and cultural landscapes.

National geoparks provide opportunities for sightseeing, vacations, leisure, health care, cultural entertainment, and scientific activities. They also play a vital role in protecting geological relics and the ecological environment, serving as hubs for geological research and education.

The Dictionary of Tourism Earth Science defines a geopark as possessing 'scientific significance, rare natural attributes, exceptional aesthetic value, and geological relic landscapes of notable scale and distribution, integrated with natural and cultural landscapes that also embody ecological, historical, and cultural importance' (Chen et al. 2013). Geoparks leverage geological relics to both safeguard these treasures and contribute to the local economy. With the aim of

Fig. 4 Yellow River Stone Forest National Geopark, Kansu Province, China (*Source* Zhang Y.)

achieving sustainable cultural and environmental development, they provide spaces of high scientific and cultural value for activities such as sightseeing, vacations, leisure, health care, education, and cultural entertainment. Fundamentally, these definitions share a consistent purpose: to protect geological heritage resources, popularize earth science knowledge, foster tourism activities, and promote local economic and social sustainability.

Unlike in many other countries, the establishment of geoparks in China has been characterized by the active, leading role of the government. As early as the 1980s, China pioneered the creation of national parks with geological themes (Chen et al. 2013). These efforts were spearheaded by government departments responsible for protecting geological relics, initially focusing on the development of single geological relic protection areas. However, the path to fully functional geoparks proved more challenging than anticipated. For instance, in 1985, some geoscientists proposed the establishment of geological parks. Yet, significant progress wasn't made until December 1999, when the Ministry of Land and Resources convened a national meeting on geological, landform, and landscape protection in Weihai, Shandong. Experts at the meeting proposed creating geological parks to preserve geological and landscape resources, a suggestion that gained the attention of ministerial government officials (Chen et al. 2015, 2019).

In response, the Ministry of Land and Resources approved the proposal in early 2000 and initiated a pilot project. Despite differing views among various ministries, the Ministry advanced the geopark initiative, establishing the National Geological Heritage (Geopark) Protection Leading Group and the National Geological Heritage (Geopark) Review Committee. It also implemented the National Geopark Application Review System and developed the Guidelines for the Establishment of National Geoparks. In September of the same year, the Notice on Applying for National Geoparks was issued, which received widespread positive responses nationwide (Fig. 4).

In March 2001, the first batch of 11 national geoparks, including Stone Forest in Yunnan Province, was approved. This was followed by the approval of 33 national geoparks, including Huangshan in Anhui Province, in February 2002. In March 2004, the third batch of 41 national geoparks, such as Wangwushan in Henan Province, was approved. Finally, in September 2005, the fourth batch of 53 national geoparks, including Mount Taishan in Shandong, received approval. This period marked a phase of rapid development for national geoparks in China. Over just four years, the Ministry of Land and Resources approved a total of 138 national geoparks. As noted, 'The promotion and establishment of National Geoparks have enabled the effective protection of China's precious geological relics, advanced geoscience research and education, stimulated local economic and social development, and supported the sustainable coordination of the economy, society, and environment'. Thus, geoparks were established not only to conserve geological heritage but also to promote geoscience education and sustainable development.

To enhance geopark management and standardize application processes, the Ministry introduced a new mechanism for granting qualifications and naming national geoparks in 2009. The Notice on Strengthening the Application and Examination and Approval Work of National Geoparks was issued that same year (MLR 2009). Since then, national geopark qualifications are granted first to ensure quality management before officially conferring the title of National Geopark. In parallel, the Ministry issued the Technical Requirements for National Geopark Planning Compilation, hosted management training programs, guided local planning revisions, and strictly enforced the national geopark planning approval process. This approach represented a landmark in the development of a geopark management model with Chinese characteristics. It also addressed previous issues of 'inaction after declaration and establishment'.

The approval process for national geoparks was refined into three stages: qualification certification, planning and construction, and acceptance and naming. This strategy helped curb tendencies to prioritize quantity over quality or focus solely on attaining designations without proper planning. With the introduction of irregular inspections, the development of geoparks entered a stable phase. As of April 2022, a total of 281 national geoparks had been approved, spanning all provinces and cities in China. This makes China the country with the largest number of geoparks worldwide. The establishment of geoparks in China is primarily intended to protect valuable geological relics. The National Main Functional Areas Plan issued by the State Council in 2010 further emphasized this goal by designating National Geoparks as restricted development zones. Corresponding management methods and regulations have ensured the effective protection and management of many of China's precious geological relics (Zhang and Jin 2021) (Fig. 5).

Fig. 5 Datong volcanic cluster National Geopark, Shanxi Province, China (*Source* Zhang Y.)

At the beginning of the 21st century, geoparks were independently conceived in China and Europe, jointly forming the foundation of the Global Geoparks Network. Guided by the philosophy of 'Celebrating Earth Heritage and Sustaining Local Communities', Global Geoparks are defined as geographical areas of international geological significance, with clearly defined boundaries and sufficient size to fulfil their functions. These areas take an integrated approach to conservation, education, research, and sustainable development, managing all resources—both natural and cultural—within their scope (UNESCO 2015).

Building on the strong foundation of National Geoparks, China's global geopark program developed rapidly after its inception in 2003. In February 2004, eight geoparks, including Huangshan, recommended by the Chinese government and successfully approved as global geoparks. That same year, the Global Geoparks Network (GGN) was established, with the first batch of members comprising above mentioned eight global geoparks and 17 European geoparks. The inaugural Global Geopark Conference, held in Beijing, highlighted China's achievements in geological heritage protection and geopark development. It underscored the Chinese government's commitment to preserving Earth's heritage, promoting science education, and advancing sustainable development.

In 2005 and 2006, an additional four and six geoparks in China, respectively, became members of the GGN. To maintain regional balance, the GGN established a limit of two global geopark nominations per country each year. As of January 2025, UNESCO had recognized 213 Global Geoparks across 48 countries in the Asia–Pacific, Europe, Africa, Latin America and the Caribbean, and North America. China has an impressive 47 global geoparks (Fig. 6).

3 Categories and Features

Currently, China's geoparks are classified into three categories: UNESCO Global Geoparks, National Geoparks, and Provincial Geoparks. UNESCO Global Geoparks are sustainable development areas evaluated and certified by UNESCO, jointly overseen by UNESCO and the Global Geopark Network. National and Provincial Geoparks, on the other hand, are natural parks authorized by the State Forestry and Grassland Administration and relevant provincial authorities. These geoparks are integrated into the management framework of the 'Natural Reserve System with Chinese Characteristics', which prioritizes national parks as the core, nature reserves as the foundation, and various natural parks as complementary components (State Council 2019).

Before 2015, Global Geoparks were approved and managed by the independent Global Geoparks Network, with

Fig. 6 Guanling fossil National Geopark, Guizhou Province, China (*Source* Zhang Y.)

their statutes, guidelines, and evaluation processes consistently supported by UNESCO. As the role of global geoparks in advancing sustainable development practices across many countries, including China, grew increasingly significant, the international community urged UNESCO to reassess its relationship with the Global Geoparks Network.

In response to these calls for change, in November 2015, during the 38th General Conference of UNESCO, all 195 member states unanimously agreed to establish the International Geoscience and Geoparks Program (IGGP). They adopted new statutes and guidelines, ensuring that after member states completed the required procedures, all global geoparks would undergo evaluation and transition into UNESCO Global Geoparks (UNESCO Global Geopark, UGGp). (See Fig. 3: UN General Assembly meeting.)

UNESCO Global Geoparks follow a comprehensive evaluation system that includes assessments for new applications as well as four-year revalidations. According to the statute, an independent evaluation team oversees the annual applications, expansions, and revalidations of global geoparks, ensuring adherence to established guidelines and submitting reports to the UNESCO Global Geoparks Council. This rigorous evaluation and revalidation process is a critical measure for fostering the high-quality development of Global Geoparks and sustaining their vitality. Moreover, it distinguishes UNESCO Global Geoparks from other UNESCO designations (Zhang and Jin 2021).

The geopark offers a vibrant array of activities, including scientific research, science popularization, tourism, heritage protection, promotion, and public outreach. Most scientific research initiatives are conducted in collaboration with professional institutions such as universities, research institutes, and survey organizations. Additionally, hosting and supporting graduate students for thesis internships has become a routine and integral part of the work carried out by various geoparks.

Danxiashan Global Geopark prioritizes the promotion of geopark branding, with a particular emphasis on the ongoing development of its 'Scientific Famous Mountain' initiatives. By training popular science volunteers and establishing key science outreach programs—China Danxia Campus and Danxiashan Nature Academy—the geopark has garnered widespread recognition. Designated as the 'Guangdong Primary and Secondary School Student Research and Practice Education Base', the geopark and its science popularization efforts have received significant acclaim. It stands as a national model for advancing nature education, research practices, and public science outreach. (See Chap. 9) (Fig. 7).

4 Roles and Challenges

After more than 20 years of planning and construction, China has developed a model of 'government-led development with extensive participation from social force'. A comprehensive system for constructing and managing geoparks at global, national, and provincial levels has been preliminarily established. This system has delivered significant benefits, including the protection of geoheritage, the promotion of earth science research and popularization, the growth of geotourism, job creation for local communities, attraction of social investment to improve infrastructure, and the advancement of international exchanges and cooperation. In recent years, China's global geoparks have played a vital role

Fig. 7 The stunning sandstone landforms of Danxiashan UNESCO Geopark make it one of the most attractive tourist destinations in China (*Source* Zhang Y.)

in inheriting local culture, promoting targeted poverty alleviation, and fostering an environmentally and ecologically conscious society. These efforts have achieved several key objectives that benefit society as a whole.

4.1 Geoconservation

Global Geoparks have effectively protected valuable geoheritage of international significance. This success is attributed to sound management practices, strengthened laws and regulations, and improved planning and infrastructure that enhance public awareness and tourist engagement. As a result, residents take pride in living within a UNESCO Global Geopark.

4.2 Scientific Research and Popularization

The development of geoparks has greatly advanced scientific research and public education. Guided by experts in geoscience, geoheritage protection, and geotourism, geoparks benefit from consulting and evaluation teams that ensure high standards. The process of establishing a global geopark involves studying and summarizing the evolution of regional geology and geomorphology. Identifying geoheritages of international value depends on rigorous scientific research. These findings, supported by geopark management, provide rich content and stories for the scientific interpretations to the public.

4.3 Geotourism

Geotourism within China's global geoparks has created numerous employment opportunities for local communities. As practitioners of sustainable development strategies, these geoparks are becoming ideal destinations for eco-friendly tourism. By 2010, geoparks in China were estimated to have generated 266,600 direct jobs and 2,154,600 indirect jobs (MLR 2012). The brand value of UNESCO Global Geoparks highlights their immense market potential.

4.4 Social and Non-social Investment

Geoparks attract investment from various sources. Government funding serves as seed capital, encouraging private and social investments to support emerging global brands.

Between 2002 and 2011, the central government invested 2.495 billion yuan in provincial geoheritage protection funds (primarily for geoparks), while local governments contributed 3.47 billion yuan. Geoparks themselves raised an additional 2.582 billion yuan, spurring related infrastructure investments of up to 18.812 billion yuan (MLR 2012). As sustainable development gains momentum, global geoparks are expected to continue fostering investments in environmental improvements and better living conditions.

4.5 International Exchanges and Cooperation

Global Geoparks promote the 'Cooperation-Sharing-Networking' mechanism. Platforms such as the Global Geopark Conference and the Asia-Pacific Global Geoparks Network Symposium enable geoparks to share management knowledge and establish partnerships, such as sister park relationships. These initiatives broaden the global perspectives of geopark managers. Annual events like Earth Day, Asia-Pacific Geopark Week, International Mountain Day, International Museum Day, and International Geodiversity Day have become regular local celebrations within China's geoparks. Young managers actively participate in nature education, science popularization, and volunteer training, which significantly enhance the influence and visibility of Global Geoparks. Moreover, UNESCO Global Geopark evaluators, including emerging professionals in geopark management, contribute to domestic and international evaluations, further strengthening the Global Geoparks Network.

4.6 Network Participation

In China, the geopark front-line managers often face challenges in independently handling daily affairs related to Global Geoparks Network, such as email correspondence, organizing and participation activities, and maintaining the GGN News Feed. Delayed feedback hampers their ability to engage promptly in common activities within the Global Geopark Network, creating the impression of a lack of enthusiasm or interest in these initiatives.

4.7 Management Capability

The management teams of some geoparks lack stability, and administrative influence is overly dominant. Collaborative management skills alone are inadequate for meeting stakeholder demands and do not align with the Global Geopark's principles of statutes, guidelines, and evaluation-based governance. Additionally, the initiative innovation is weak, with geopark-specific creative activities falling short of the potential presented by China's extensive family of geoparks.

4.8 Top-Down Mentality

The government-led approach to geopark management has resulted in limited community participation. Local communities are often excluded from decision-making processes, and their aspirations as stakeholders are not adequately represented. Greater emphasis should be placed on adopting a bottom-up approach to ensure that community representatives' voices are heard before major decisions are made. This shift will help prevent future challenges, facilitate smooth policy implementation, and strengthen sustainable development and cultural preservation. Engaging the community will also foster a sense of pride and belonging, reinforcing their involvement in geopark development.

4.9 International Relationships

There is an imbalance in the development of domestic and international partnerships. While 'friendship geoparks' within the country are thriving, international 'sister geoparks' remain relatively inactive. China's global geoparks lack a strong presence in international conferences and GGN activities, with limited participation and interaction. Language barriers and rigid, uniform regulations contribute to these issues, highlighting the need for urgent improvements.

4.10 Operational Ineffectiveness

The 'cooperative partner' mechanism has been instrumental in advancing scientific research, nature education, and capacity building, as well as in fostering talent development and team building. However, some geopark management teams rely excessively on external partnerships. These teams often lack familiarity with UNESCO Global Geopark statutes and guidelines, leading them to delegate responsibilities, such as evaluations and revalidations, to third parties or external consultants. This practice violates Criterion III of UNESCO Global Geoparks, which requires that geoparks establish a legally recognized management agency staffed with multi-disciplinary professionals capable of overseeing the entire geopark. Such shortcuts hinder the growth of local management teams and undermine the long-term development of geoparks.

5 Conclusions

Geoparks are specially designated areas that showcase the geological wonders of our natural environment. At the provincial, national, and global levels, they emphasize the vital role that geological and geographical features play in creating a better planet for all. UNESCO Global Geoparks stand out as the benchmark for excellence, embodying adherence to rigorous standards and objectives. Their influence continues to grow, captivating public interest and setting new precedents for sustainable development. Chinese geotourism and geopark development has a long history to share. There are still many challenges, both old and new, to overcome. By sharing each other stories through interacting and learning with other countries, they could possibly be turned into positive opportunities of bringing benefits to all people. Danxiashan, established as one of the first global geoparks in 2004, serves as a model for success in this respect. It reflects a journey of overcoming challenges to achieve its goals, fostering a healthier environment and a richer community.

References

Chen et al (2013) Dictionary of tourism earth science. Science Press, Beijing, p 506

Chen A, Lu YT, Ng Y (2015) The principles of geotourism. Springer, Berlin, p 264

Chen A, Ng Y, Zhang E, Tian M (2019) Dictionary of geotourism. Springer, Singapore, p 786p

Eder W (1999) 'UNESCO geoparks'—a new initiative for protection and sustainable development of the Earth's heritage. N Jb Geol Paläont Abh 214(1/2):353–358

Ji S, Pan J (1997) The geological heritage sites and national parks in China. In: Engineering geology and the environment, Mannos, Köukis, Balkema, pp 2961–2965

Jones C (2008) History of geoparks. Geol Soc Lond Spec Publ 300:273–277

Ministry of Land & Resources (MLR) (1995) Regulations on the protection and management of geological relics (Minister Order No. 21), Article 8. Beijing

Ministry of Land & Resources (MLR) (2009) The notice on strengthening the application and examination and approval work of national geoparks. Notice N0. 50. Beijing

Ministry of Land & Resources (MLR) (2012) Report on geoparks development in China, Beijing, 128p

State Council (2019) Guiding opinions on establishing a protected natural areas system with National Parks as the mainstay. General Office, Beijing, June 2019

UNESCO (2015) Operational guidelines for UNESCO global geoparks, 10p. https://unesdoc.unesco.org/ark:/48223/pf0000391354. Accessed 19 Jan 2025

UNESCO (2024) The UNESCO global geoparks book: celebrating earth heritage, sustaining local communities. https://www.unesco.org/en/articles/unesco-global-geoparks-book-celebrating-earth-heritage-sustaining-local-communities. Accessed 01 Dec 2024

UNESCO (2025) General information, definitions, governance and framing issues. https://www.unesco.org/sites/default/files/medias/fichiers/2023/01/uggp_application_process_faqs.pdf?hub=67817. Accessed 20 Dec 2024

Wimbledon WAP et al (1998) A first attempt at a geosites framework for Europe—an IUGS initiative to support recognition of world heritage and European geodiversity. Geol Balc 28(3–4):5–32

Wimbledon WAP et al (1999) Geological world heritage: geosites—a global comparative site inventory to enable prioritisation for conservation. Mem Descr Carta Geol D'it LIV 1999:45–60

Zhang Z (1997) Proceedings of the 30th international geological congress, Beijing, China, 4–14 August 1996: geosciences and human survival, environment, natural hazards - global change, vols 2 and 3. CRC Press, 364p

Zhang J, Jin X (2021) A review of the evaluation system of UNESCO global geoparks. Geol Rev 67(06):1763–1769

Zhang J. & Zhang J. (2022). The history and outcomes of the 30th International Geological Congress, Beijing (1996). Institute for the History of Natural Science, CAS, Beijing. https://www.iugs.org/_files/ugd/f1fc07_2b4ce7148c584fcda741dcd9cb157e21.pdf?index=true

Zhao X, Zhao T (xxx) National geoparks initiated in China: putting geoscience in the service of society. Episodes 26(4):302–309. https://pdfs.semanticscholar.org/8412/eda385ccf90a657fe3c12bf07c341dc77aac.pdf

Geology

Liuqin Chen, Fusheng Guo, and Fujun Liu

Abstract

The Danxia Basin, in northern Guangdong Province, is one of continental fault basins in the late Mesozoic Basin-and-Range Tectonic System in Southeast China. The Cretaceous basin filling is characterized by red-colored siliciclastic rocks (red beds) with a thickness of about 4000 m. These red beds are nearly horizontal and vary from pebbles to clay in grain size. The texture, sedimentary structures, lithology, and fossils are collectively indicative of a superaqueous oxidization environment. Among them, the Upper Cretaceous Danxia Formation is subdivided into three members, namely Bazhai, Jinshiyan, and Baizhaiding Member in ascending order. The Bazhai and Baizhaiding members are dominated by alluvial conglomerates, while the Jinshiyan Member is featured by eolian cross-bedded sandstones. They are hard to support the well-developed cliffy slopes in the basin. The cutting by intersecting fractures have controlled the spatial assemblage of mesas, cliffs, and canyons. Cavernous weathering forms are well developed on sandstone and conglomerate walls. The landscape and landforms have had a pivotal and prolonged impact on cultural relics including ancient temples, villages, and cliff inscriptions. The outstanding geodiversity and geoheritage serve to make Danxiashan an ideal site for geotourism and geoscience education.

L. Chen (✉) · F. Guo · F. Liu
School of Earth and Planetary Sciences, East China University of Technology, Nanchang, Jiangxi, China
e-mail: chenliuqin@ecut.edu.cn

1 Tectonic Setting

At the end of the Triassic Period, the collision between Indosinian and South China plates had come to an end with the regional stress transformed from compression to extension. In the background of crustal extension, a multitude of EW-trending faults were formed. These included the Qitianling-Zhuguangshan, Dadongshan-Guidong, Huashan-Guposhan and Lianyang-Fogang Faults. Some of these faults were filled with the Indosinian granites. A lot of pre-existing faults were reactivated and filled by large amounts of granites during the Yanshanian orogenic cycle (Deng et al. 2011). A basin and range tectonic system was formed during late Mesozoic in South China (Wang and Shu 2012). The ranges were composed of volcanic and intrusive granitic complex, whereas the adjacent basins were mostly medium- to small-sized fault basins filled with reddish terrestrial clastic rocks (Fig. 1).

The basement of the Danxia Basin is composed of a compound syncline of Carboniferous strata. The syncline is characterized by an NNE-trending axis, steep SE limb, a gentle NW limb, and a NW trending axial plane (Wu et al. 2001). The Shaoguan-Renhua Fault, a northern branch of the Wuchuan-Sihui Fault, is a transgressional fault crossing the middle of the basin. There are also several NNE-trending faults parallel to the major fault. The western limb thrusts above the eastern one. These faults are generally long in length and deeply cut with long and active duration. The fault activity began before the Indo-sinian period, which formed several Cretaceous basins. Shen (2014) identifies 11 major faults, with lengths between 6 and 20 km and widths from 20 to 50 m. These faults are primarily transgressional.

Y. Ng and W. Eder (eds.), *The Redstone Park of China*, Geoheritage, Geoparks and Geotourism,
https://doi.org/10.1007/978-3-031-95603-4_3

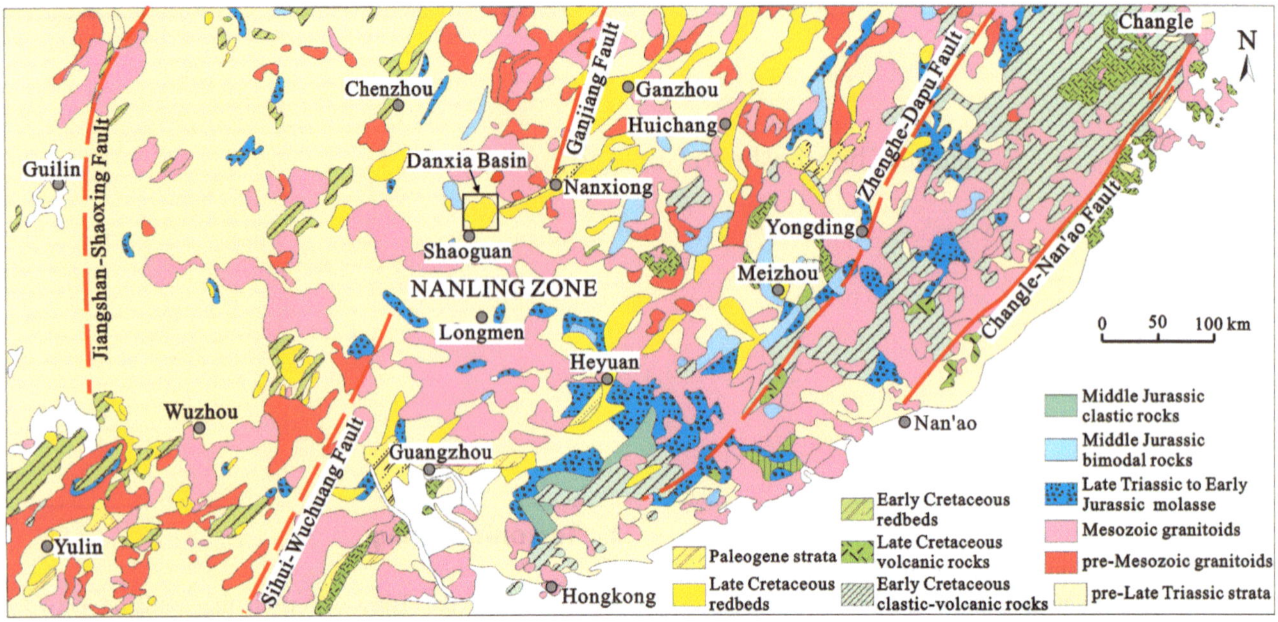

Fig. 1 Geological map showing the dominant faults and Mesozoic-Cenozoic fault basins and volcanic-intrusive complex in Southeast China (after Shu et al. 2009)

2 Stratigraphy

The Danxia Basin of northern Guangdong Province is a small-sized fault basin and covers an area of about 500 km^2 (Figs. 2 and 3). The thickness of Cretaceous red beds is close to 4,000 m, among which the Upper Cretaceous red beds are the most widespread ones.

The Danxiashan area belongs to the stratigraphic division of northern Guangdong Province of South China. The strata at the basin margin are composed of Cambrian, Ordovician, Devonian, Carboniferous, Permian, Triassic and Jurassic systems in an ascending order (Table 1).

The study of the red beds in the Danxiashan area can be traced as early as the nineteenth century. These red beds were called as deck sandstone in the book *'China'* written by the well-known German geologist Richthofen Ferdinand von (1833–1905) who went by Guangdong Province in 1869 and 1870 during his field trips in China. During the 1920s and 1930s pioneer geologists such as Feng and Zhu (1928), Chan (1938), and Chen and Liu (1939) performed stratigraphic division and correlation of the Danxia Basin.

From the 1950s on, the research on the red beds of the Danxia Basin has increased particularly on their formation ages and biostratigraphy. Based on paleontology, paleomagnetism and isotope dating data, the red beds of the Danxia Basin were classified into Changba Formation (Early to Middle Cretaceous) and Danxia Formation (Late Cretaceous) by Zhou (1990) and Xu et al. (1990). Using ostracoda fossils, Zhang (1992) further divided the red beds of the basin into Sandong and Maziping formations (early to middle Early Cretaceous), Changba Formation (late Early Cretaceous to early Late Cretaceous), and Danxia Formation (middle to late Late Cretaceous).

As shown in Table 2, the Changba Formation is further divided into four members, and it is at present considered to be deposited during Late Cretaceous based on ostracoda fossils (Zhang and Lin 2013). Specifically, the second member of the Changba Formation is featured by ostracoda fauna *Mongolocypris-Cypridea-Darwinula* assemblage, indicating they were deposited during the Cenomanian Stage, while the fourth member is rich in ostracoda fauna *Cypridea-Altanicypris-Candona* assemblage, suggesting a depositional age of Turonian Stage of Late Cretaceous.

The first and third members of the Changba Formation are dominated by yellowish-brown conglomerate and coarse-grained sandstone with interbedded fine-grained sandstone and siltstone, whereas the second and fourth members are largely composed of purple red and greyish brown fine-grained sandstone, muddy siltstone with small amounts of conglomerate and sandstone, and are rich in fossils and calcareous paleosols (Fig. 4). The Changba Formation is less resistant to weathering owing to the large proportion of fine-grained components.

At the Changba section (Fig. 5), the first member of the Changba Formation is 480 m thick and composed dominantly of brick red conglomerate and pebbly sandstone, underlain unconformably by light greyish white quartz sandstone of Early Carboniferous Ceshui Formation (Zhang 1992). The second member is about 850 m thick and consists of yellowish-brown pebbly fine sandstone, purplish red silty

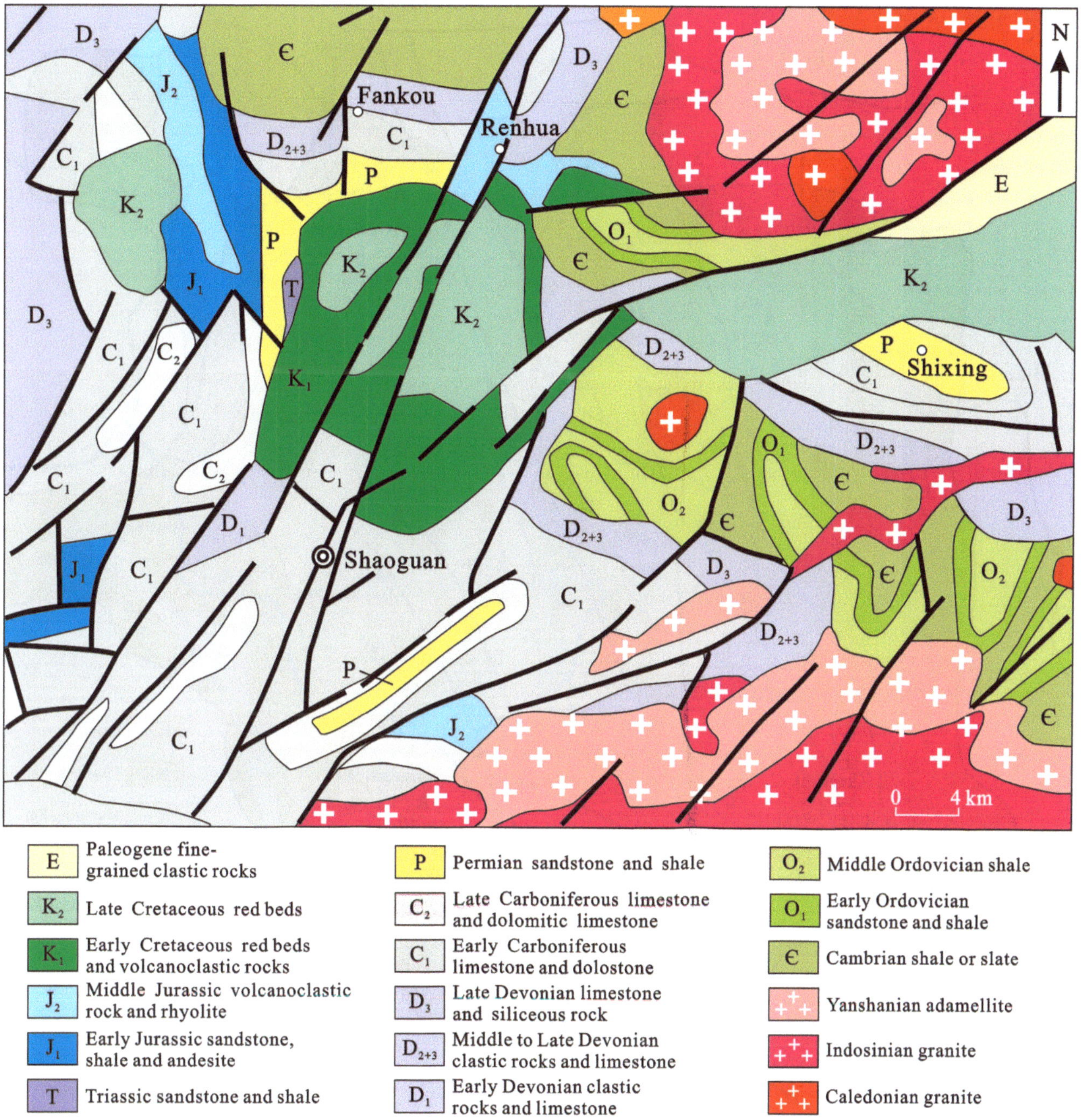

Fig. 2 Geological map of the Danxia Basin and adjacent areas (Cai et al. 2002)

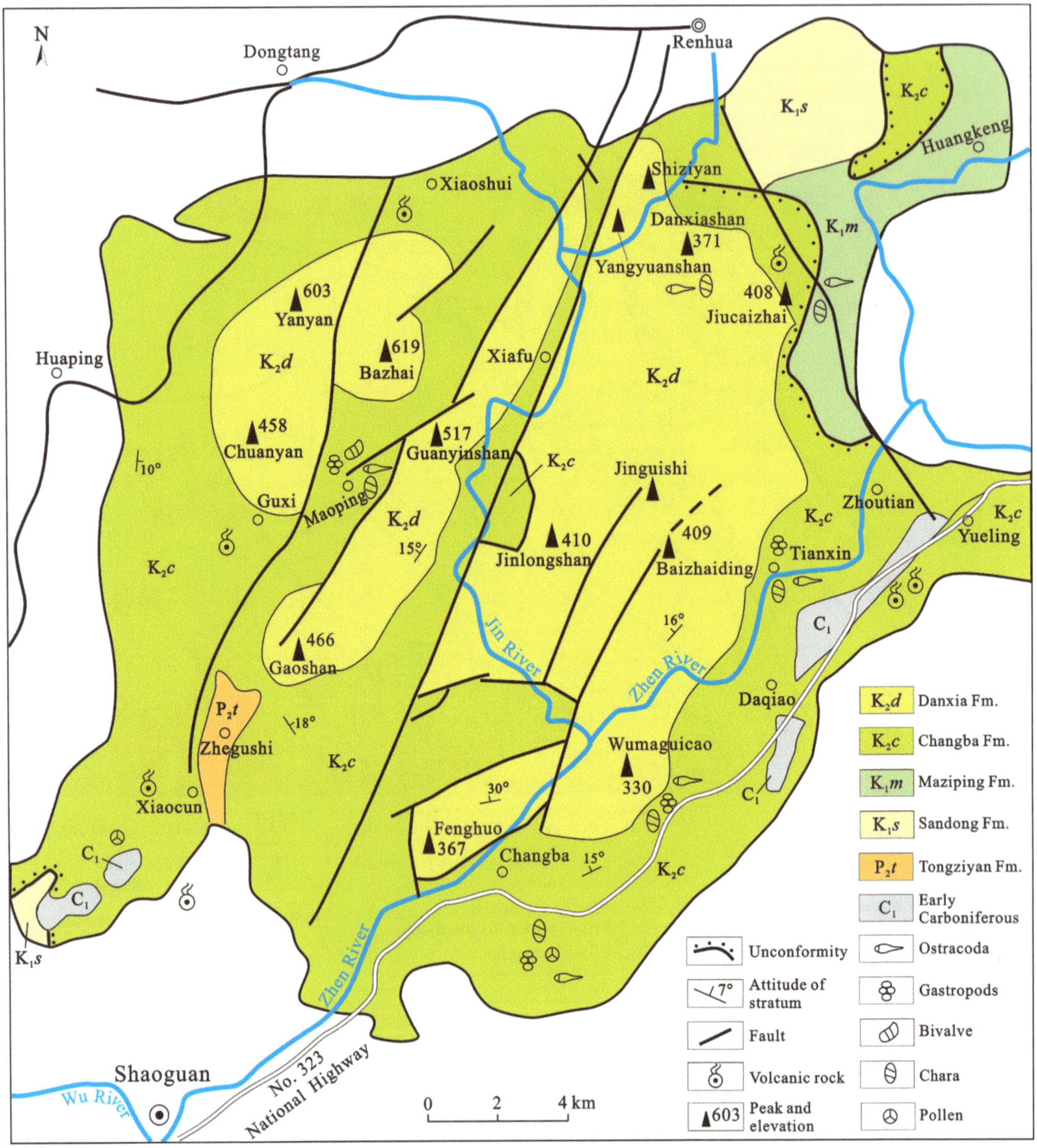

Fig. 3 Geologic map of the Danxia Basin (modified from Zhang and Lin 2013)

Table 1 Summary of stratigraphy at the outer edge of the Danxia Basin (compiled from Wu 1994)

Erathem	System	Series	Group (formation)	Code	Lithology and depositional environment	Thickness (m)
Mesozoic	Jurassic	Middle	Maziping Gr	J_2m	Interbedded fluvial and volcanic rocks	>1,040
		Lower	Qiaoyuan Fm	J_1q	Feldspar quartz sandstone, siltstone, and mudstone	380–500
			Jinji Fm	J_1j	Alternating continental and marine coal-bearing deposits	273–329
	Triassic	Lower	Daye Fm	T_1d	Shallow marine carbonate rocks	>500
Upper Paleozoic	Permian	Upper	Changxing Fm	P_2c	Shallow marine carbonate rocks	16–97
			Longtan Fm	P_2l	Alternating continental and marine coal-bearing deposits	115–118
					Shallow marine sandy and muddy siliciclastic rocks	60–115
		Lower	Tongziyan Fm	P_1t	Alternating continental and marine coal-bearing deposits	274–340
			Wenbishan Fm	P_1w	Shallow marine muddy siliciclastic rocks	165–195
			Qixia Fm	P_1q	Shallow marine carbonate rocks	40–140
	Carboniferous	Upper	Chuanshan Fm	C_3c	Shallow marine carbonate rocks	58–325
		Middle	Huanglong Fm	C_2h	Shallow marine carbonate rocks	115–500
		Lower	Zimenqiao Fm	C_1z	Shallow marine carbonate rocks	80–162
			Ceshui Fm	C_1c	Alternating continental and marine coal-bearing deposits	196–300
			Shidengzi Fm	C_1s	Shallow marine carbonate rocks	280
			Liujiatang Fm	C_1l	Shallow marine carbonate rocks	100
			Menggong'ao Fm	C_1m	Shallow marine sandy and muddy rocks	350–470
	Devonian	Upper	Maozifeng Fm	D_3m	Coastal to shallow marine sandy and muddy rocks	200–250
			Tianziling Fm	D_3t	Shallow marine and tidal flat carbonate rocks	167–826
		Middle	Huanggangling Fm	D_2h	Coastal to shallow marine muddy carbonate rocks	100–250
		Lower-Middle	Laohutou Fm	$D_{1-2}l$	Coastal to shallow marine sandy and muddy siliciclastic rocks	120–400
			Yangxi Fm	$D_{1-2}y$	Fluvial sedimentary rocks	>200
Lower Paleozoic	Ordovician	Middle-Upper	Changkengshui Fm. And Longtouzhai Gr	O_{2-3}	Marine carbonate and siliciclastic rocks	>929
		Lower	Xinchang Fm. and Xiahuangkeng Fm	O_1	Marine siliciclastic rocks Interbedded volcanic rocks	164–268
	Cambrian		Bacun Gr	Є	Grey to greyish-green low-graded metamorphic flysch formation	>3,830

Table 2 Stratigraphic division and lithology of the Danxia Basin (Compiled from Wu 1994)

System	Series	Formation	Member	Code	Brief lithologic description
Cretaceous	Upper	Danxia	Baizhaiding	K_2d^3	Brownish red conglomerates, pebbly sandstones, medium to coarse grained sandstones and siltstones. The thickness is generally 40–60 m with the maximum of 224 m because of the top erosion
			Jinshiyan	K_2d^2	Dominant fleshy red to maroon thick- to massive-bedded arkose sandstone with large-scaled cross-beddings. Small amounts of silty mudstone, granule conglomerates, pebbly sandstones. Thickness is 92–300 m
			Bazhai	K_2d^1	Maroon massive conglomerates and pebbly sandstones with local purple red thin-bedded silty mudstone. Thickness is 194–570 m
		Changba	Fourth	K_2c^4	Purple red silty mudstone and muddy siltstones with small amounts of interbedded fine-grained sandstones. The top is dominated by brownish red and brick red thick-bedded arkose sandstone. Thickness is 164–680 m
			Third	K_2c^3	Purple red conglomerates and pebbly sandstones. Thickness is 342–545 m
			Second	K_2c^2	Purple red thick to thin-bedded silty mudstone and muddy siltstones with medium- to thin-bedded fine-grained sandstones and siltstones. Thickness can reach 1,001 m
			First	K_2c^1	Purple red conglomerates and pebbly sandstones. Thickness is 483 m
	Lower	Maziping Fm		K_1m	Purple red silty mudstone and muddy siltstones with small amounts of fine-grained sandstones. The top is dominated by brownish red and brick red thick-bedded arkose sandstone. Thickness is 164–680 m
		Sandong Fm		K_1s	Dark purple red volcanic rocks and volcaniclastic rocks overlain unconformably on the Jurassic Qiaoquan Formation. The lower is dominated by Rhyolite tuff, andesite, basalt, and the upper is tuff, tuff sandstone, purple red siltstone, muddy siltstones. Thickness is 54–128 m

mudstone and muddy siltstone with abundant paleosols. The third member is about 300 m thick and composed of grayish- to yellowish-brown conglomerate, coarse sandstone with interbedded siltstone. The fourth member is about 360 m thick overlain conformably by the Danxia Formation, which is dominated by brownish red silty mudstone and muddy siltstone with small amounts of fine to medium-grained sandstone and well development of paleosols.

The Danxia Formation is divided into three members, namely Bazhai, Jinshiyan, and Baizhaiding Member in ascending order. The Bazhai and Baizhaiding members are conglomerate-dominated and were deposited in alluvial fan systems, while the Jinshiyan Member is characterized by eolian sandstones with large-scale cross-beddings (Fig. 6).

Almost all escarpments developed in the Danxia Formation are located within the geopark boundary. It is well exposed at the Zhanglaofeng and Bazhai sections with clear sedimentary sequences and structures. At the Zhanglaofeng section, the conglomerate-dominated Bazhai Member can be observed from the road to the bottom of the Jinshiyan Temple, while the Jinshiyan Member is characterized by aeolian sandstones with large-scaled cross-beddings. The uppermost of the Danxia Formation with the beginning of the Baizhaiding Member, is recognized by the increase of conglomerates.

At the Zhanglaofeng section, the Bazhai Member of the Danxia Formation is about 490 m thick, the Jinshiyan Member is 125 m thick, and the Baizhaiding Member is only 40 m thick (Fig. 7). Escarpments, mesas, and peaks are well developed in the Bazhai and Jinshiyan members. The Bazhai Member is mainly outcropped in the east and north of the basin, which is dominated with brownish red and purple conglomerate and pebbly sandstone with the presence of ostracoda and stonewort fossils. The Jinshiyan Member is distributed within the east and middle of the basin, which is readily distinguished by aeolian sandstones with large-scaled cross-beddings and impressive inscriptions on cliffs. The Baizhaiding Member is dominated by brownish red conglomerate and sandstone with stonewort fossils (Zhang 1992).

Fig. 4 Representative rocks and sedimentary structures of the Changba Formation. **a** Massive purplish mudstone beds intercalated with fine-grained sandstone beds of the second member; **b** thick sandstone beds interacted with thin mudstone beds of the third member; **c** close-up of the interbedded sandstone and mudstone beds; **d** fine-grained sandstones with bioturbation structures

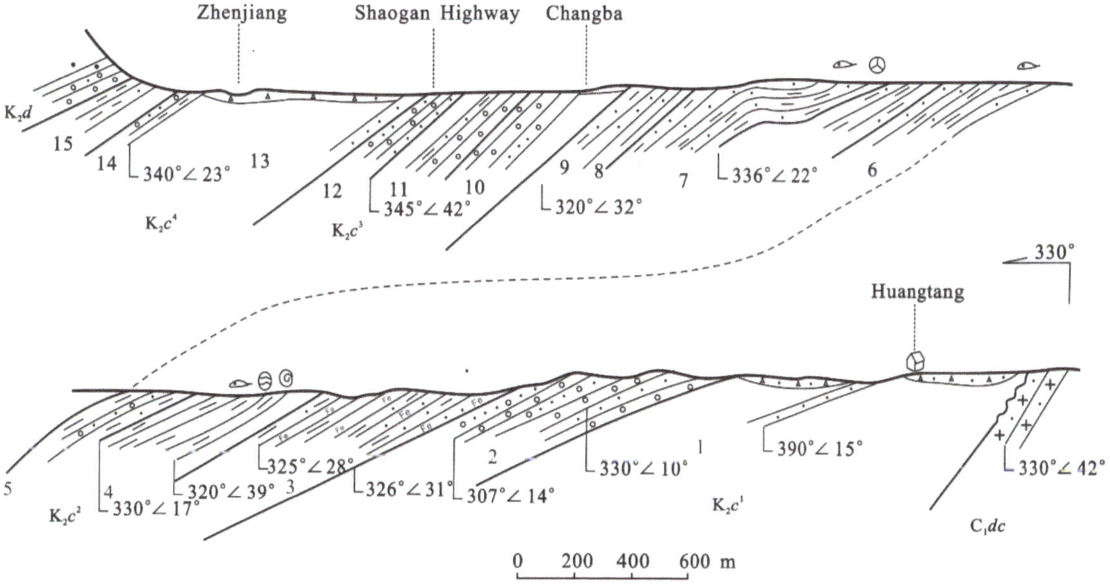

Fig. 5 The Changba section from Huangtang village to northern Zhen River (Zhang 1992)

Fig. 6 Representative rocks and sedimentary structures of the Danxia Formation. **a** Massive conglomerates of the Bazhai Member; **b** parallel-bedded conglomerates of the Bazhai Member; **c** Eolian sandstones with large-scale cross beddings of the Jinshiyan Member; **d** Interbedded fine-grained sandstone and mudstone beds with well-developed mud cracks of the Jinshiyan Member

3 Sedimentary Rocks

3.1 Conglomerate

Conglomerates are very common in Mount Danxiashan, which have supported the development of prominent cliffs of danxia landscape (Fig. 8a). The lower (Bazhai Member) and upper (Baizhaiding Member) parts of the Danxia Formation are mainly composed of conglomerates (Fig. 8b). They are thinly bedded in the sandstone dominated Jinshiyan Member. The conglomerates are clast-supported, subrounded and poorly sorted (Fig. 8c). Pebbles are composed of quartz, quartz sandstones, limestones, granites, and small amounts of low-grade metamorphic rocks and andesite. The largest pebble (quartz sandstone from the Bazhai Member at the Lingxi section) found in the field is 30 cm in diameter. They are massive, parallel-bedded, and cross-bedded with small amounts of sandstone intercalations. In general, basal erosion surfaces develop at the bottom of conglomerate beds, and imbrication structures are very common. These observations indicate the conglomerates were mostly deposited in alluvial fans (Chen et al. 2019a, b).

At the Changba section, Goushitan near the Zhen River, the fine pebbly conglomerates are normally graded with reddish angular mud pebbles, indicating strong river erosion of the lower floodplain sediments (Fig. 8d). The thickness of a single graded bedding can vary from a few to several tens of centimeters, although some may reach more than 100 cm.

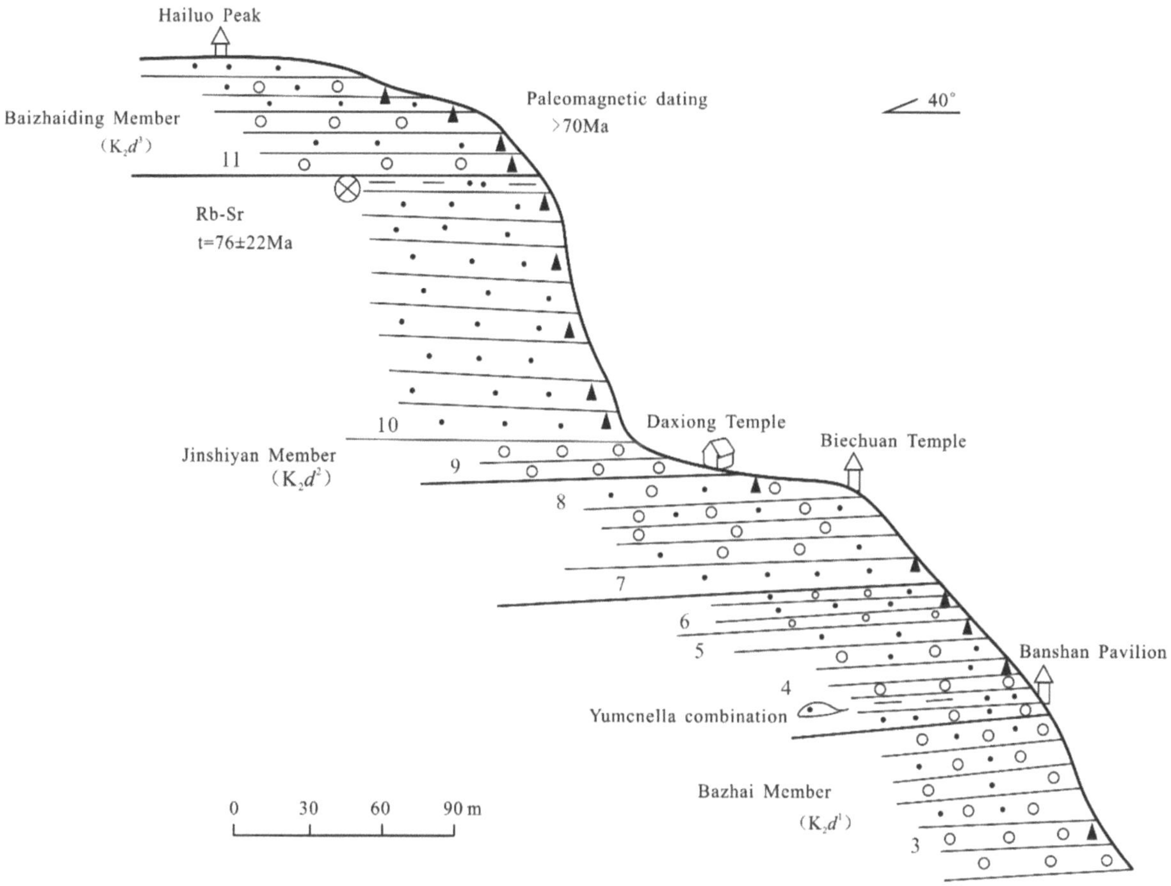

Fig. 7 Sketch profile of the Danxia Formation at the Zhanglaofeng section (Zhang 1992)

3.2 Sandstone

Sandstone beds are generally interbedded with conglomerate and mudstone beds. They may have sharp boundaries and display gradual grain-size changes. They are usually found with abundant sedimentary structures including cross, parallel and horizontal beddings, ripple marks, and trace fossils. The horizontal bedding is planar, laying parallel to each other on the plane surface. The thickness of each lamina is mostly below 1 mm with occasional thickness of 1–2 mm.

It is notable that the middle part (Jinshiyan Member) is exclusively composed of eolian sandstones with large-scale cross-beddings (Fig. 9). The thick and large-scale cross beddings of aeolian sandstones can reach tens of meters, and the dip angles of their foresets are usually greater than 30°.

The eolian sandstones are highly textural mature with sub-rounded to rounded detrital grains, so the cliffs are smoothed and were preferably chosen for ancient rock carvings. This explains why most cliff inscriptions of Mount Danxiashan are located on the eolian sandstone cliffs, and many of them became accredited as National Key Cultural Relics Protection Sites in May 2013.

Under a polarized or scanning electronic microscope, the detrital particles consist mainly of quartz, feldspar and various rock fragments, with a small amount of heavy minerals and other materials. The sand grains are well-rounded, well-sorted, and cemented by calcite.

At the Jinshiyan Temple of Zhanglaofeng Peak and Luyitang of Shuimeifen (Sleeping Beauty) Peak, honeycomb weathering features are well-developed on the eolian sandstone cliffs (Chen et al. 2022a, b).

Fig. 8 Conglomerates of the Danxiashan formation. **a** Overview of a rough cliff supported by conglomerates of the Bazhai Member; **b** closeup view of the conglomerates in photo **a**; **c** subangular pebbles of a conglomerate intercalation in an abandoned sandstone quarry at Lingxi River; **d** Conglomerate bed with erosional base and reddish mudstone pebbles

3.3 Siltstone and Mudstone

The fine-grained clastic rocks are dark red or fuchsia and composed of siltstones and mudstone. They are generally interbedded with sandstone beds with abundant mud cracks and trace fossils.

Mud cracks, also known as shrinkage/desiccation cracks, are a kind of surface sedimentary structures formed by dehydration and shrinkage of sediments because of extreme drying conditions. The width of these cracks is generally a few millimeters to a few centimeters, extending in a broken line or curve. They are often connected in a T or Y-shape and divide the plane into a series of polygons. In the vertical profile, the cracks are usually wedge-shaped and filled by the overlying sandy sediments, and they are often sausage-shaped owing to compaction during diagenesis (Fig. 10). A polygonal network of mud cracks can be seen on the bottom surface of the overlying conglomerate unit.

A trace fossil, also called an ichnofossil, indicates ancient biological activities on the surface of or in sediments, including footprints, migrating traces, burrows, boreholes, and other structures such as egg fossils and feacal fossils. Based on common ecological habits, trace fossils include residential, crawling, anchorage (rest), foraging, feeding, cultivation and escape trails (EBSCO 2024; Wikipedia 2024).

Fig. 9 Representative sandstones and sedimentary structures. **a** and **b** Large cross beddings in eolian sandstones of the Jinshiyan Member of the Danxia Formation; **c** asymmetrical ripple marks on the fine-grained sandstone surfaces of Danxia Formation; **d** trace fossils in the fine-grained sandstones of the Chagnba Formation

In the Danxiashan Global Geopark, trace fossils are mainly found in the fine-grained sandstone and mudstone of the second and fourth members of the Changba Formation and the Jinshiyan Member of the Danxia Formation. In the red mudstone of the Jinshiyan Member, the columnar burrows *Scoyenia* are perpendicular or oblique to the bedding surface, with a diameter of 5–10 mm, and they are often associated with mud cracks.

4 Bedrock and Fracture Control on Danxia Landscape

4.1 Cliffs Formed by Steep to Vertical Fractures

A series of steep to vertical joints and faults were formed in the basin, along which canyons were generated owing to water erosion and collapse of the sides. At Zhanglaofeng Peak, there is a typical normal fault with a displacement of 40 cm (Fig. 11), which can be clearly recognized by the boundary between the conglomerate and sandstone beds. It is a slot canyon with only one person passing by at the narrowest section. Other similar canyons include Fuyinxia and Tianmenshan.

When the canyons widened, escarpments started to form. The Jinshiyan Cliff is a representative landform in the study area. Rockfalls were produced because the lower soft rock beds were easily weathered or eroded. There is a huge rockfall at the Jinbifujin pool (Fig. 12), where the rockfall is composed predominantly of conglomerates and the lower part is composed of interbedded sandstones and mudstone. Owing to lower permeability, the water seeps out along the lower part of the cliff to form pools near the tourist path.

4.2 Cliffs Supported by Conglomerates and Sandstones

The conglomerates are harder than sandstones or other finer-grained clastic rocks in the study area. This had

Fig. 10 Mudstone beds and mud cracks. **a** Interbedded sandstones and mudstones of Changba Formation; **b** close view of the mudstone beds; **c** massive development of mud cracks in the Jinshiyan Member of Danxia Formation; **d** mud cracking impression at the bottom of a conglomerate bed

been confirmed by rebound measurements using a Schmidt Hammer (Guo et al. 2023). The underlying Changba Formation of Danxiashan is characterized by interbedded sandstones and mudstones or paleosols. It is different to the Danxia Formation because of its high mud content and relatively low compressive strength, making it less resistant to weathering. The cliffs are mainly supported by harder conglomerates and sandstones of the Danxia Formation (Fig. 13). The red beds are nearly horizontal or slightly inclined. Cavities were formed along bedding planes due to differential weathering and are called bedding-controlled cavities (Peng et al. 2015; Guo et al. 2020).

Mount Yangyuanshan is a huge EW-trending rock wall with a length of 710 m and a width of 140 m (Fig. 14). The famous Yangyuanshi Rock is a rock column located at the eastern end of the Yangyuanshan, with a height of 28 m and a diameter of 7 m. The corresponding one at the western end is called Hunyuanshi Rock, which is a core stone. Therefore, Yangyuanshan is a tableland of red beds dissected by EW-trending and SN-trending joints and subsequently forming the peculiar Yangyuanshi column by weathering.

The conglomerates and sandstones are porous and permeable, providing pathways for water and salt movement which favor salt weathering (Chen et al. 2021). The rocks are generally rich in feldspar minerals and are less resistant to weathering. Except for the aeolian sandstones of the Jinshiyan Member, the conglomerates of the Danxia Formation are low in textural maturity with poorly sorted subangular to subrounded clasts. Significant amounts of calcite and iron oxide cement are present in the sandstones (Peng et al. 2014; Chen et al. 2019a, b) and conglomerates (Chen et al. 2021). The average pH value of rainwater is 4.41 while the interbed water in the area is 7.18. There are slight increases in the ion contents of K^+, Na^+, Mg^{2+}, Cl^-, NO_3^-, SO_4^{2-}, and significant increases in Ca^{2+} and HCO_3^- in the interbed water samples (Peng et al. 2014). The increases in pH values and ion contents of Ca^{2+} and HCO_3^- are the result of chemical reactions between acid rainwater and carbonate minerals. At the same time, the feldspar minerals were transformed into clay minerals by hydrolysis. Therefore, the dissolution weathering of these chemically unstable components have resulted in the disintegration of the bedrock.

Fig. 11 Slot canyon was formed along a normal fault

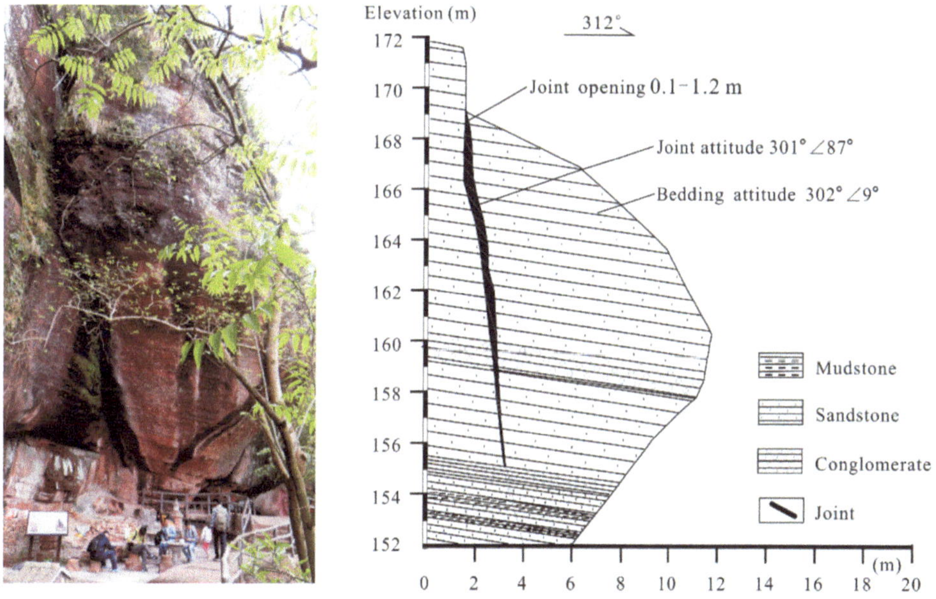

Fig. 12 Rockfall was formed by steep joints and erosion of lower fine-grained sedimentary beds

Fig. 13 Cliff is supported by interbedded conglomerates and sandstones (Chen et al 2022a). **a** Chuantoushi Rock of the Zhanglaofeng Peak **b** sketch showing clear boundaries between sandstones and conglomerate beds. A red circle marks a person as a scale

5 Conclusions

The Danxia Basin is one of fault basins formed during late Mesozoic time in a crustal extension setting in Southeast China. The basin filling is characterized by red-colored clastic rocks that were deposited in alluvial fans, river, lake, and desert environments. In particular, the Danxia Formation of Late Cretaceous age is dominated by well-developed thick-bedded conglomerates that support cliffy escarpments. The middle part of the Danxia Formation, Jinshiyan Member, is featured by eolian sandstones with large-scale cross-beddings, rounded to well-rounded and well-sorted sand grains. Most cliff inscriptions are preserved along the eolian sandstone cliffs owing to its high textual maturity. Therefore, it was deposited in a desert environment that was probably related to a prolonged period of aridity during the Campanian Stage of the Cretaceous Period in low-altitude areas of East Asia.

Danxia landscape is featured by red coloration and well-developed cliffy escarpments. The red coloration was formed owing to thin films developed around detrital sand grains. The thin films are composed of iron oxides which were formed by oxidation in depositional environments. The steep slopes, cliffs, and canyons were produced owing to weathering and erosion along fractured systems. Caverns of different sizes ranging from centimeters to meters in diameter is another distinctive landform of danxia landscape. In general, the cavernous weathering features in conglomerates are larger than those in sandstones.

The detailed studies carried out in Mount Danxiashan have provided a better picture of the formation and characteristics of the appealing geomorphological features of danxia landscape. These studies are important in geodiversity and geoheritage assessments for geoconservation and sustainable tourism development. They also serve to make Mount Danxiashan an ideal site for conducting geoscience education and science popularization activities. Global comparative studies with similar landscapes in the United States and Australia are encouraged, as they will be valuable for the development of long-term conservation and development plans for Mount Danxiashan and other Danxia landscapes in China.

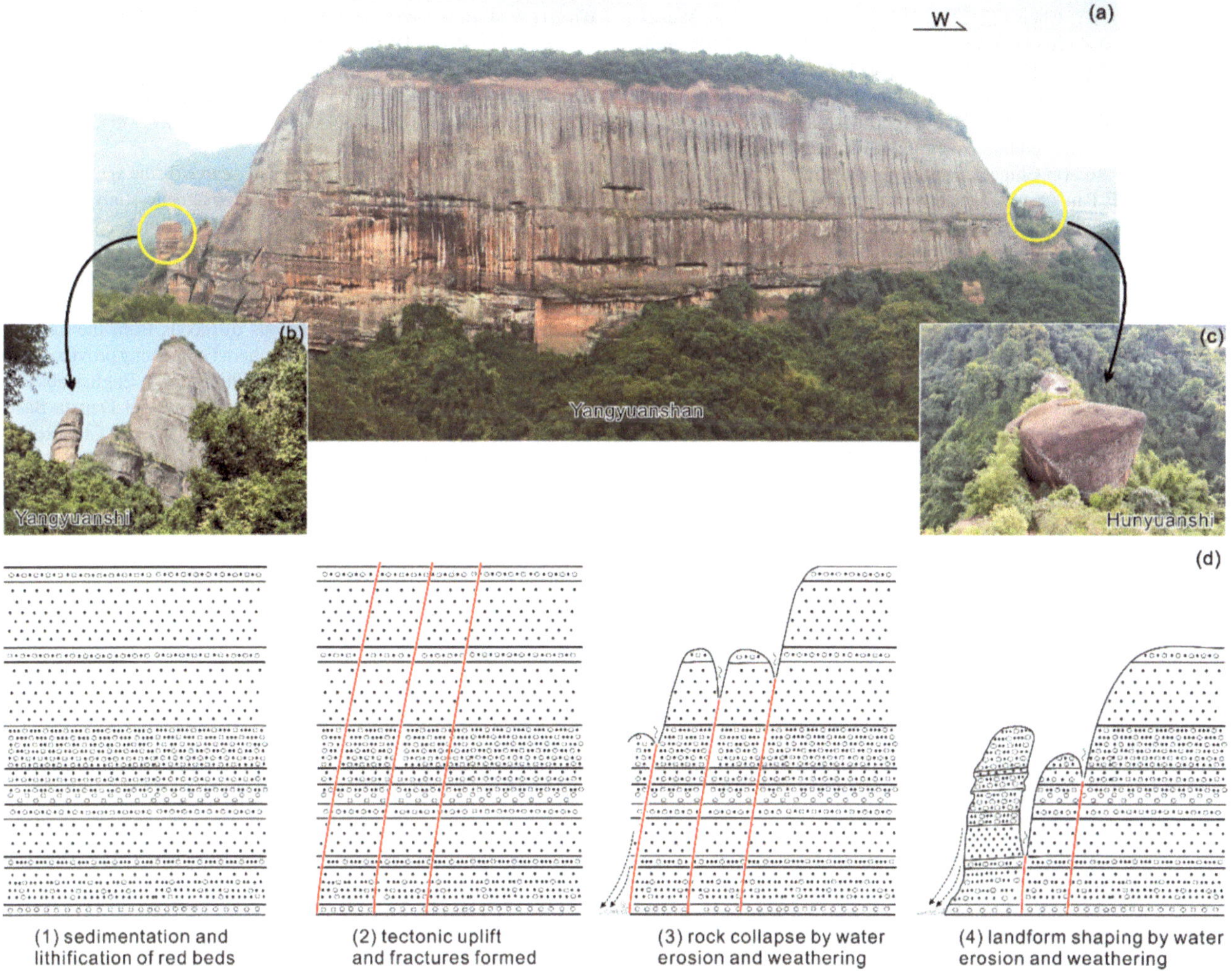

Fig. 14 Formation of the Yangyuanshi pillar (Chen et al 2022a). **a** The huge Yangyuanshan rock wall extends from east to west with an approximate length of 600 m. **b** The famous Yangyuanshi pillar at the eastern end; **c** The Hunyuanshi column at the western side; **d** A development model of the Yangyuanshi pillar

References

Cai SK, Lu J, Huang ZH, Li RW, Li ZM (2002) Geology of Guangdong province, Hong Kong and Macao special administrative regions. In: Ma, L.F. Geological Atlas of China. Beijing: Geological Publishing House. 253–260 (in Chinese)

Chan KT (1938) On the subdivisions of the red beds of south-eastern China. Bull Geol Soc China 18(3–4):301–324

Chen GD, Liu HS (1939) Geology of the Gongshui River Valley in Jiangxi Province. Jiangxi Geol Surv (Number 2):164 (Fig. 3 Fig. C Illustrates) (in Chinese with English abstract)

Chen LQ, Guo FS, Liu FJ, Xu H, Ding T, Liu X (2019a) Origin of tafoni in the Late Cretaceous aeolian sandstones, Danxiashan UNESCO Global Geopark, South China. Acta Geol Sin-Engl Ed 93(2):205–217

Chen LQ, Li PC, Guo FS, Liu X, Li XM (2019b) Facies analysis and paleoclimate implications of the Late Cretaceous Danxia Formation in the Danxia Basin, Northern Guangdong Province, South China. Acta Sedimentol Sin 37(1):17–29 (in Chinese with English abstract)

Chen LQ, Guo FS, Shao CJ, Du DD, Chen F, Luo M (2022a) Geodiversity characterization of the Danxiashan UNESCO Global Geopark of China. Int J Geoheritage Parks 10(4):459–476

Chen LQ, Shi YX, Ng Y, Guo FS, Chen X, Hu HP, Peng SL (2022b) Roles of algae in honeycomb weathering in humid climate: an example from Danxiashan UNESCO Global Geopark of China. Geoheritage 14(4):120

Chen X, Chen LQ, Zhang YH, Du DD, Hu HP, Liu DX, Li WZ (2021) Lithological and environmental controls on large tafoni along conglomerate cliffs in subtropic humid Danxiashan UNESCO Global Geopark. J Mt Sci 18(5):1131–1143

Deng P, Ren JS, Ling HF, Shen WZ, Sun LQ, Zhu B, Tan ZZ (2011) Yanshanian granite batholiths of southern Zhuguang Mountain: SHRIMP zircon U-Pb dating and tectonic implications. Geol Rev 57(6):881–888 (in Chinese with English abstract)

EBSCO (2024). Trace fossils. https://www.ebsco.com/research-starter/environmental-sciences/trace-fossils. Accessed 23 Oct 2024

Feng JL, Zhu HS (1928) Geology and Mineral Resources in Qujiang, Renhua, Shixing, Nanxiong, Guangdong province. The Annual Report of Geological Survey in Guangdong and Guangxi (number 1) (in Chinese)

Guo FS, Chen LQ, Yan ZB, Liu FJ, Pan ZX, Zhang WQ, Hu HP (2020) Definition, classification, and danxianization of Danxia landscapes. Acta Geol Sin 94(2):361–374 (in Chinese with English abstract)

Guo FS, Ling YY, Chen LQ, Zhou WP, Li HW, Cheng LK, Wu ZC, Li GR, Guo Z (2023) Controlling factors and types of geomorphologic landscapes in Danxiashan UNESCO Global Geopark of China. Geoscience 37(6):1665–1679 (in Chinese with English abstract)

Peng H, Qiu ZW, Pan ZX (2014) Experimental study on the weathering features of bedding caves at Mt. Danxiashan. Sci Geogr Sin 34(4):454–463 (in Chinese with English abstract)

Peng H, Ren F, Pan ZX (2015) A review of Danxia landforms in China. Z Geomorphol (Supplementary Issue 1):19–33

Shen P (2014) The evolution mirrored by the Cretaceous sedimentary sequences in the Nanxiong-Danxia Basin and its response to the tectonic transition at the SCS's area. Doctoral dissertation, China University of Geosciences, Beijing (in Chinese with English abstract)

Shu LS, Zhou XM, Deng P, Wang B, Jiang SY, Yu JH, Zhao XX (2009) Mesozoic tectonic evolution of the Southeast China Block: New insights from basin analysis. J Asian Earth Sci 34(3):376–391

Wang DZ, Shu LS (2012) Late Mesozoic basin and range tectonics and related magmatism in Southeast China. Geosci Front 3(2):109–124

Wikipedia.org (2024) Trace Fossil. https://en.wikipedia.org/wiki/Trace_fossil. Accessed 24 Oct 2024

Wu QJ (1994) Basic geological characteristics of Danxia Basin. Proceedings of the Second Danxia Landscape Tourism Development Symposium, 1–21 (in Chinese)

Wu JT, Liu JX, Liao ST (2001) Geological characteristics, evolution of the Danxia Basin. Reg Geol China 20(3):274–279 (in Chinese with English abstract)

Xu X, Ge, TM, Xiao H (1990) Preliminary search on the age for red beds in Danxia Basin-new evidence of paleomagnetic study. Journal of Chengdu College of Geology, 17(2):76–86 (in Chinese with English abstract)

Zhang XQ (1992) Cretaceous subdivision and correlation in the Danxia Basin. J Stratigr 16(2):81–95 (in Chinese)

Zhang XQ, Lin XY (2013) Cretaceous ostracods from the Changba formation in the Danxia Basin of northern Guangdong province. Acta Micropalaeontol Sin 30(1):58–86 (in Chinese with English abstract)

Zhou HJ (1990) Classification and age of the strata of Danxia Basin, Guangdong province. Guangdong Geol 5(4):51–58 (in Chinese with English abstract)

Geomorphology

Ke Zhang

Abstract

The 'Danxia landscape' has been studied for nearly a century, ever since Professor Feng Jinglan first proposed the concept of 'Danxia layer' and described its associated landforms in 1927. The way to define 'Danxia landscape' has long been much discussed, based on its rock colors, geomorphic shapes, depositional structures, ancient depositional environment, cementing materials and geological structures that shaped the landforms. More than twenty different definitions have so far been proposed (Wang et al. 2020; Peng 2020a, b; Guo et al. 2020).

1 Definition

The 'Danxia landscape' has been studied for nearly a century, ever since Professor Feng Jinglan first proposed the concept of 'Danxia layer' and described its associated landforms in 1927. The way to define 'Danxia landscape' has long been much discussed, based on its rock colors, geomorphic shapes, depositional structures, ancient depositional environment, cementing materials and geological structures that shaped the landforms. More than twenty different definitions have so far been proposed (Wang et al. 2020; Peng 2020a, b; Guo et al. 2020).

'Danxia landscape' is composed of spectacular reddish sandstone landforms. Thus, the reddish color is a key defining criterion in describing 'Danxia landscape'. Nevertheless, reddish could be in a wide range, from brownish yellow, brown, purplish red, brownish red, brick-red, to grayish purple. The color problem will be discussed specifically in the following section. Second, the 'Danxia landscape' is composed of relatively less resistant or softer rocks with high geomorphic sensitivity (Allison 1993). They are relatively softer than the more resistant and harder rocks such as silica cemented quartzose sandstones. They, therefore, can be easily sculpted by exogenic processes of weathering, wind blowing, flowing water and gravity effect especially those containing hard and coarse fragments but weak cements of calcareous, ferruginous, or argillaceous (clay) cements, particularly in pseudo-karst formation. The morphology of a 'Danxia landscape' usually appears with flat top, steep cliffs, and gentle foothills (Fig. 1). It is closely related to nearly horizontal rock layers and relic planation surfaces, and often associated with vertical joints, faults and collapsed colluvium in pediment. The red beds are less resistant, forming lowlands or geomorphic basin like Mount Danxiashan area as compared to the surrounding higher highlands and mountains composed of more resistant rocks such as quartzose sandstone, granite, and gneiss. The diversified shapes of landforms in Danxiashan area are a result of million years of denudation and large-scale differential weathering. The gentle foothill gradient is often controlled by internal friction angle.

Third, high escarpments were commonly developed on the relatively thick layers of conglomerate, sandy conglomerate, and sandstone red beds which have relatively weak cement, particularly the thick or very thick beds, composed of conglomerate, gritstone, and sandstone with uniform depositional structure. These blocky structures, in humid and warm area like South China easily collapse under the effect of vertical joint cutting or lateral erosion of rivers (Peng 2020a, b; Fig. 2), forming distinctive reddish cliffs or hill blocks (Fig. 1). In addition, those thick red beds overlying on the weak rock layers such as thin siltstone or mudstone layers are susceptible to mass movement and rock falls. This often happens

The original version of the chapter has been revised. A correction to this chapter can be found at
https://doi.org/10.1007/978-3-031-95603-4_10

K. Zhang (✉)
Sun Yat-sen University, Guangzhou, Guangdong, China
e-mail: eeszke@mail.sysu.edu.cn

Y. Ng and W. Eder (eds.), *The Redstone Park of China*, Geoheritage, Geoparks and Geotourism,
https://doi.org/10.1007/978-3-031-95603-4_4

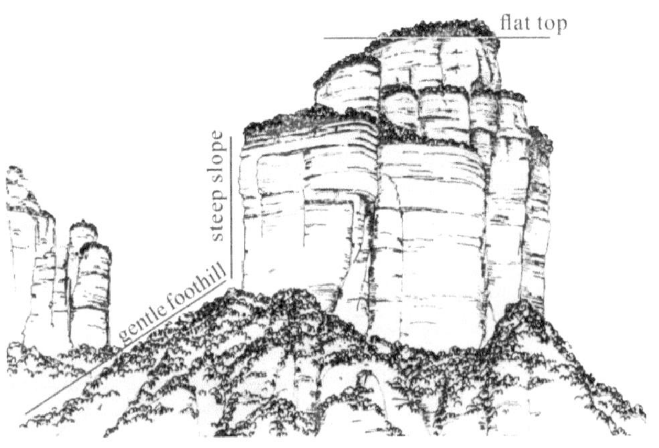

Fig. 1 Typical Danxia landscape with flat top, steep cliffs, and gentle foothills. Bazhai castle-like peak in Mount Danxiashan area (Modified after Peng 2020b)

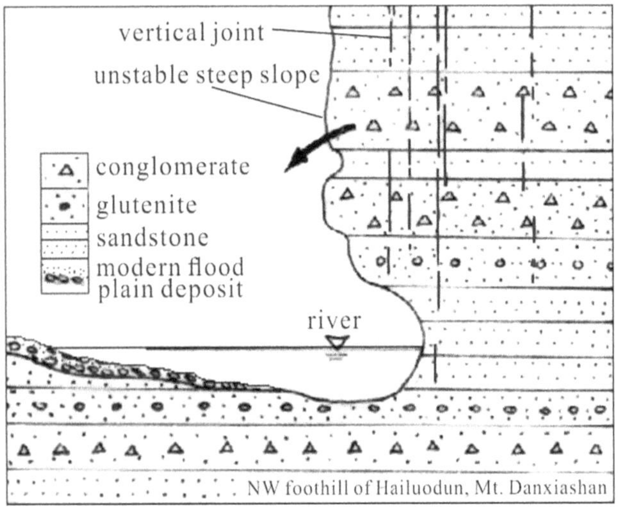

Fig. 2 Formation of steep slope by river lateral erosion and vertical joint cutting (Modified after Peng 2000)

in the cliff formation of Danxia landscape (Peng 2020a, b; Fig. 3). A good example in Mount Danxiashan, is the relatively hard rock beds of Danxia Formation, overlain on the relatively weak rock beds of Changba Formation. Added to this are the regional faults and large vertical joints cuttings which have dissected and broken the red beds, resulting in orientation of hill block trends (Fig. 4).

Finally, neotectonic uplift and tilting are prerequisite for formation of Danxia landscape, otherwise, the red beds would be underneath the land surface, or just represent as a peneplain, not the current hilly form of landscape. Periodical uplifts, or rejuvenations can produce multilayered landform of which will be later discussed in this chapter.

Mesozoic continental clastic red beds mainly composed of conglomerate and sandstone are poorly compacted and favorable for the formation of Danxia landscape. This is also a common phenomenon globally following a kind of governing law (Philips 2007). Other red beds older than Mesozoic age and non-continental origins were well compacted, and often harder, darker red and seldom formed typical topography like what have been witnessed in Danxia landscape.

All in all, the Danxia landscape comprises red beds and reddish sandstone landforms of continental origin and characterized by visualized elegant bright of sunset and sunrise colors, with rocks of Mesozoic age. The landscape is characterized by flat tops, steep cliffs, and gentle foothills, caused by geological structural controls, neotectonic uplift and exogenic processes. Outstanding examples of similar landscapes can also be observed in the USA, Peru, and Australia (DCCEEW 2025; Geologyin 2025; IUGS 2025; USGS 2025).

2 Geological Backgrounds

Mount Danxiashan is situated in the 'Danxia Basin', which is a gentle broad dual-syncline basin, striking approximately north–south with Danxia Formation (K_2) in the core and surrounded by Changba (K_{1-2}), Maziping and Sandong (K_1) Formations (Fig. 5). Geologically, Mount Danxiashan area is a Cretaceous basin but geomorphologically, it can only be regarded as a relic basin due to the long-term differential denudation since Neogene. Therefore, quotation marks for 'Danxia Basin' have been used here.

Shandong Formation (K_1) is composed of dark purple volcanic rocks and pyroclastic rocks. The upper part is composed of tuff, tuff sandstone, purplish siltstone, muddy siltstone interbedding; while the lower part is composed of rhyolitic tuff, and basalt. The total thickness can be up to ca. 128 m (Figs. 5, 6).

Maziping Formation (K_1) is a dark purplish clastic deposition, overlying conformably on Shandong Formation. The upper part is composed of dark purplish mudstone, silty mudstone intercalated with siltstone and sandstone; while the lower part is composed of dark purplish, variegated fluvial-lacustrine coarse to middle sandstone, siltstone, and silty mudstone, with the bottom layers of andesitic breccia tuff and andesitic porphyrite. The total thickness is 645–994 m (Figs. 5, 6).

Changba Formation (K_{1-2}) is composed of purple-red fluvial-lacustrine mud-siltstone, fine sandstone, and a little conglomerate with the thickness of about 1,300 m, overlying conformably on Maziping Formation (Figs. 3, 5, 6).

Danxia Formation (K_2) is composed of brick-red alluvial-fluvial conglomerate, gritstone, sandstone interbedded with thin siltstone with the total thickness of about 1,300 m, overlying conformably on Changba Formation. Danxia Formation is harder than Changba Formation in

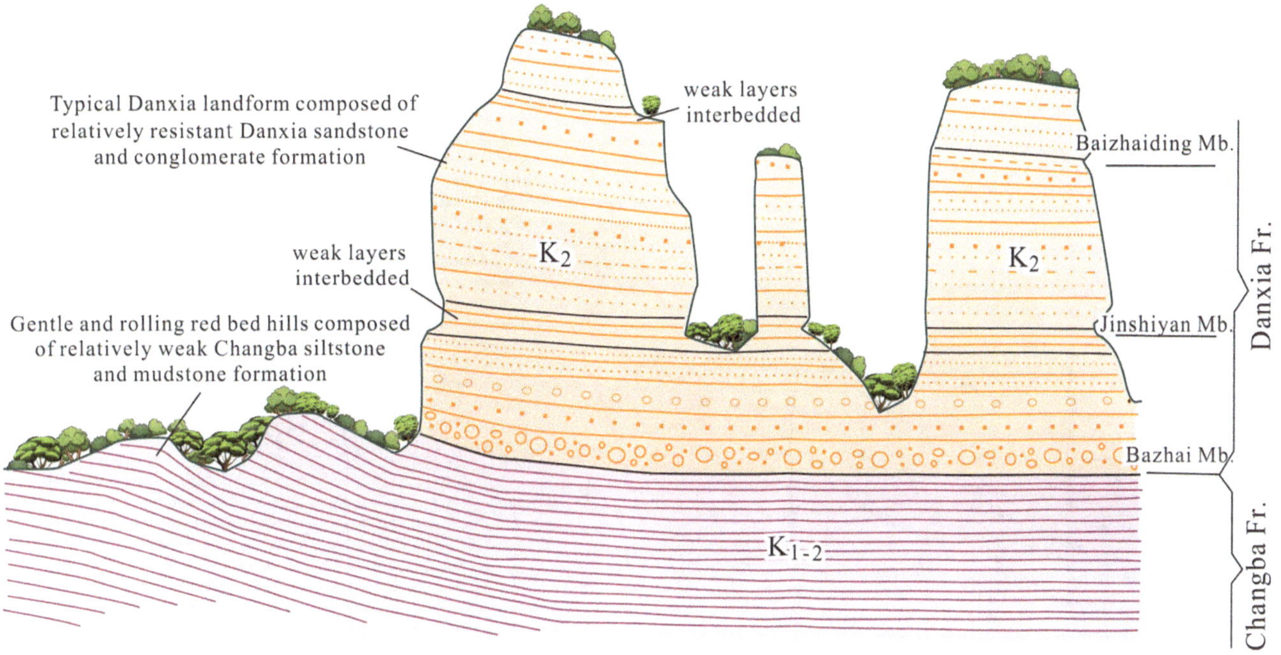

Fig. 3 Relatively hard and thick red beds (Danxia Fr.) overly on weak and thin red beds (Changba Fr.), which is favourable for the formation steep cliffs (modified after Peng 2020b)

resistance and makes up the main part of Danxia landform (Figs. 3, 5, 6). Danxia Formation can be subdivided into: Bazhai, Jingshiyan and Beizhaiding members from bottom to top (Figs. 3, 5, 6).

The regional Shaoguan-Renhua fault striking NNE direction cut across the 'Danxia Basin', dividing the basin into eastern and western parts. Besides the regional fault, there are mainly two groups of regional joints, NNE and nearly EW vertical joints, which cut into the landform and control the hill and valley trending (Figs. 4, 5).

The strata of the 'Danxia Basin' overly unconformably on Jurassic layers, rocks, and others older. The 'Danxia Basin' is different from the eastern adjacent the Nanxiong Basin with equivalent strata to Changba, and Danxia formations and with addition of early Neogene strata in that fault contacts in relationship with other older strata or rocks instead of depositional contacts (unconformity), implying that the Danxia Basin is basically a relic Proter-depositional basin instead of fault-block basin just as the Nanxiong Basin (Fig. 7).

3 Basic Morphology

3.1 Macro-Landform

Geological structure line pattern is a basic factor that controls the distribution patterns and shapes of landforms in Danxia landscape, where large tectonic lines such as faults control the general arrangement of hill blocks, while small structures such as regional joints control the trends, density, and the plane pattern of hill blocks. For instance, the arrangement of hill blocks at Mount Danxiashan extends in an NNE direction along the major fault (Fig. 4). The trend of hill blocks and the arrangement of stone pillars were mainly along secondary faults and large joints of the near EW trend (Peng 2020a, b; Fig. 4). The endogenic and exogenic processes developed various mega-landforms such as the over 200 m high steep cliffs (Fig. 8), castle-like peaks (Fig. 1), isolated peak, mesa (Figs. 9, 10), the typical 500 m long, 300 m wide and 230 high stone walls (Figs. 9, 10) (Peng 2020a, b), cuesta (Fig. 11), pyramidal peaks (Fig. 12), round-top peaks (Fig. 13), columns (Fig. 14, 15), valleys (Fig. 16), scattered peak forests (Figs. 17, 22, 23), ravines (Chinese called them 'one line sky') (Fig. 18), and stone arch (Fig. 19), caves, both vertical and large horizontal caves in group (Figs. 20, 21).

As for the formation of grooves and stratiform caves, based on the rock uniaxial mechanical experiments over the red beds in South China including those taken in Mount Danxiashan, the strength of the conglomerates is universally higher than the sandstones because of the presence of relatively more resistant fragments such as metamorphic rocks, volcanic tuff lavas, quartz and less calcite cements than the sandstones (Zhu et al. 2010). This results in differential weathering along the weak layers which leads to groove development especially under the environments of acid rain, frequent alternations of intensive freezing and thawing in the Quaternary cold-warm climate fluctuations (Zhu et al. 2010). However, it is not excluded

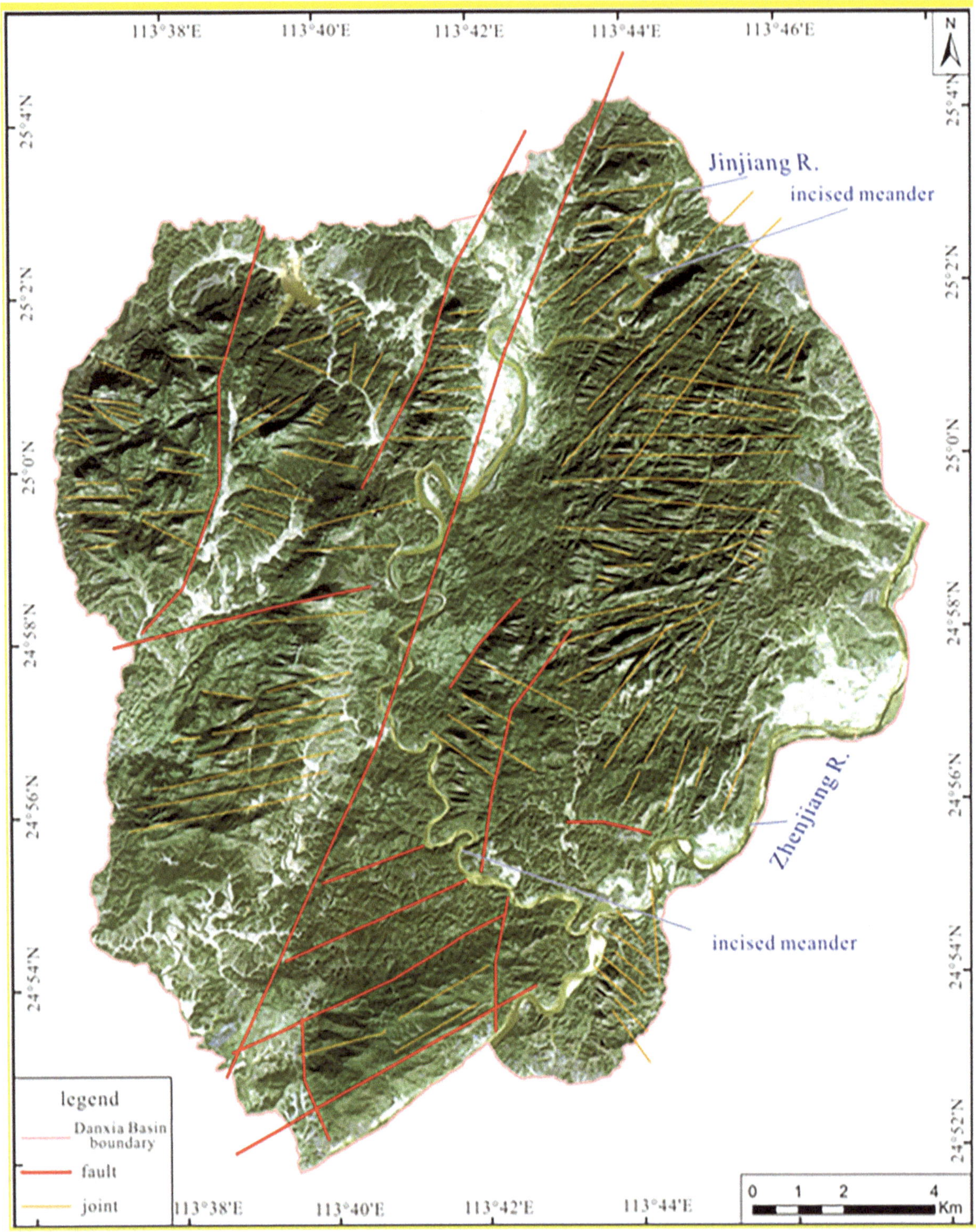

Fig. 4 High resolution satellite image of 'Danxia Basin' in Guangdong Province, showing the controls of regional faults and joints in relation to the trending of the hills, valleys and some river sections. *Note* Most faults and joints were interpreted by image line structures combining with geological map

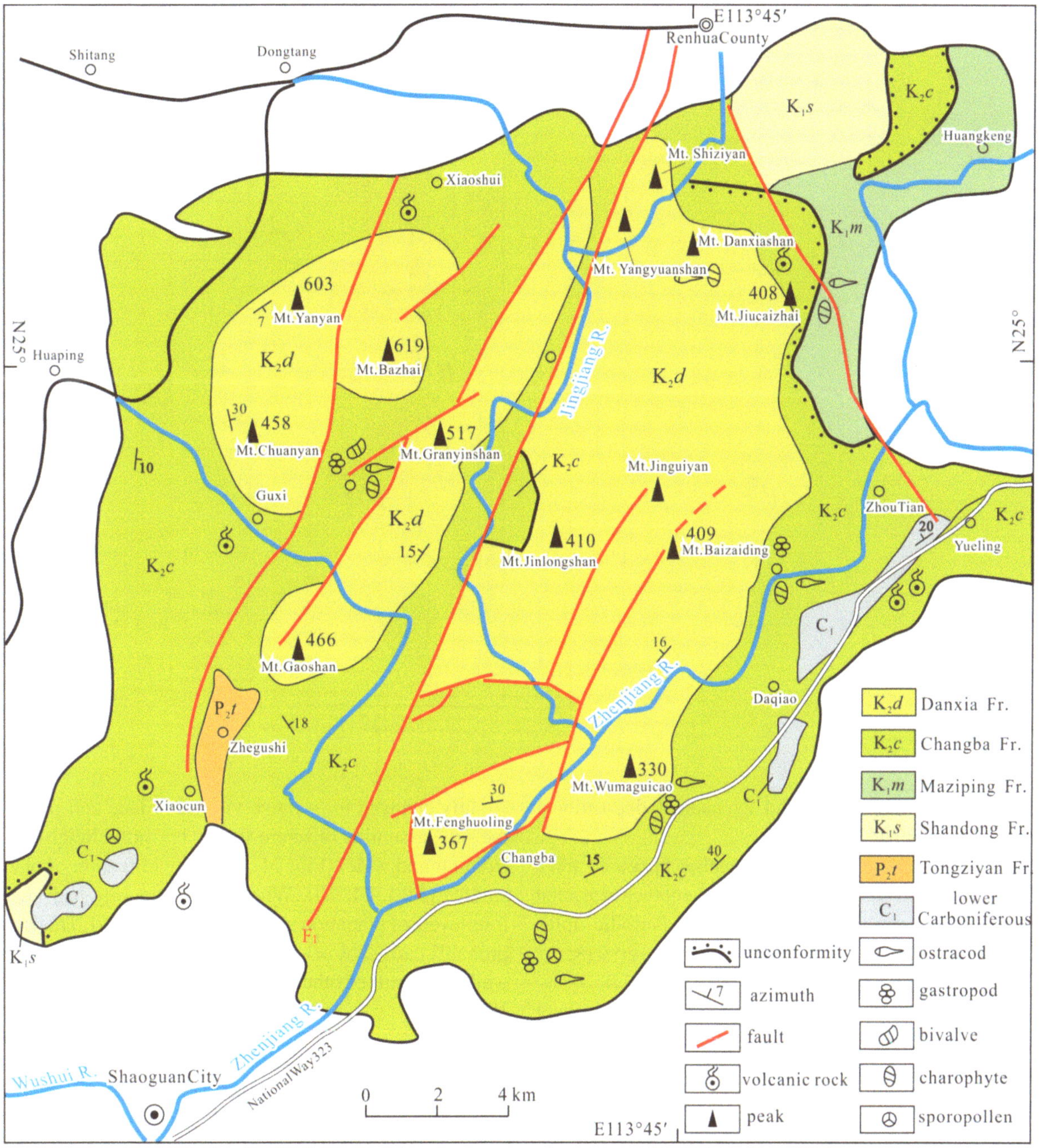

Fig. 5 Simplified geological map of the 'Danxia Basin'. *Note* The NNE major fault is Shaogran-Renhua regional fault (F_1)

from the origin of early ancient river lateral erosion and later tectonic uplifts, river down-cutting due to similar heights and layer-like distribution ignoring different lithologies (Figs. 19, 20, 21).

Neotectonic uplift also played a significant role in the formation of different landforms in Mount Danxiashan. Based on the model of the erosional cycle of Davis (1909), the evolution of Danxia landforms can be described as infant, mature and old stages. According to the theory, Danxia landscape in China had been designated in the World heritage List as China Danxia based on such age difference. The Danxia landforms of Chishui can be classified as early or infant stage, while those of Taining have been regarded as late infant stage while the Danxia landforms of Langshan belong to the early mature stage. The landforms of Danxiashan, on the other hand, belong to the middle mature stage with Longhushan belong to the late mature stages while those of Jianglangshan belong to the old stage. Even among the landforms of Danxiashan, evolution stages can be used to describe the variations, depending on different locations and

Fig. 6 Stratigraphic column of the 'Danxia Basin'

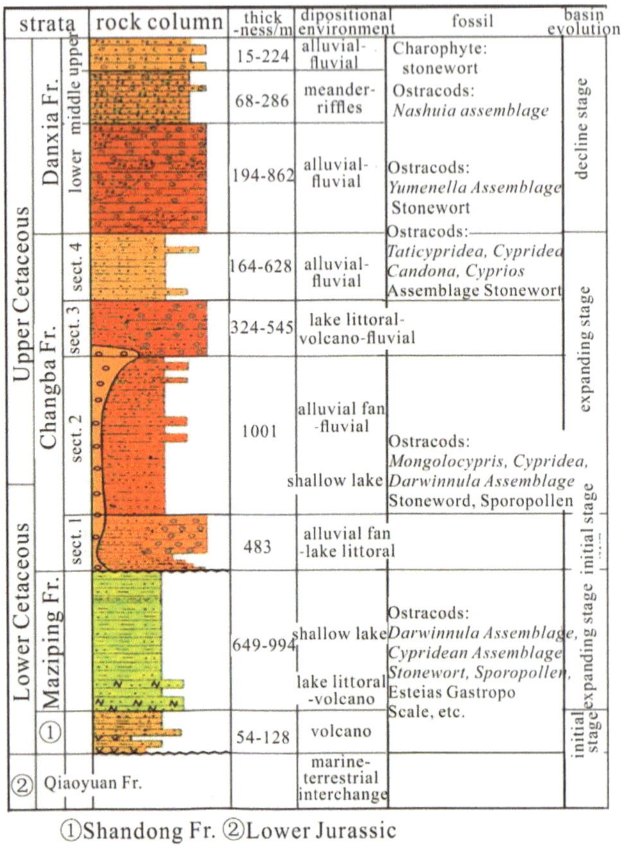

lithologies which were controlled by ancient depositional facies (Guo et al. 2020; Fig. 23).

During long old stage process, a peneplain appeared and a new uplift, i.e., rejuvenation, could trigger another erosional cycle, resulting in multilevel of landforms. Periodic uplift could produce several levels of relic planation surfaces. As early as 1938, Chen (1939) had described the relic peneplain or planation surface, in Mount Danxiashan. In his paper of Subdivisions of the Red Beds of Southeastern China, he wrote 'The younger series (of the red beds), on the contrary, frequently forms flat-topped hills with steep slopes. This is due to the gentle inclination of horizontality of the strata. Such examples are found in the Ren Hua, Qu Jiang, and Le Chang and certain districts of Guangdong where hills of about the same altitude form a nearly straight skyline when viewed from a suitable elevation'.

Chen had distinguished at least one planation surface at that time. However, later research had discovered at least two levels of planation surfaces in Mount Danxiashan, one is located at about 600 m level represented by Bazhai, and the other is about 300 m level represented by most of the hills in the area. Su Shi, an ancient poet of Song Dynasty (1037–1101 A.D.) described it as 'a range viewed in face and peaks from the side, assuming different shapes and heights viewed from far and wide' (translated by Xu Yuanchong). To describe planation surface landform, the poem could be slightly changed to 'a range viewed in face and peaks from the side, assuming different shapes but **same heights** viewed from far and wide' (Fig. 24). It is a vivid feature of planation surface (Figs. 24, 25). Whereas, Huang Jing had identified four levels of planation surfaces (Huang 2010). This could be probably caused by several differential uplifting and tilting, as well as a result of other false distinctions (Zhang and Huang 1995).

3.2 Micro-landform

Affected by depositional structures and long-term exogenic processes, niche-like caves, honeycomb-like caves, potholes, are developed as micro-landforms. Tafoni are very common on Mount Danxiashan. Lithology has been considered as the most significant controlling factor of tafoni development (Chen et al. 2019). It is enhanced by the presence of lamina siltstone and mudstone for lateral weathering and expedited by frequent rains and wind erosion. Acid dissolution produced by organisms in the moisture microclimate in curvature of the cliff walls also causes exfoliation of detrital grains and cement materials from the rock surface and encourages tafoni formation. This resembles the self-organized behavior of the dissipative structure theory. The famous geosite of 'Dragon Scale Rock' in Jinshiyan Temple of the Zhanglaofeng Scenic

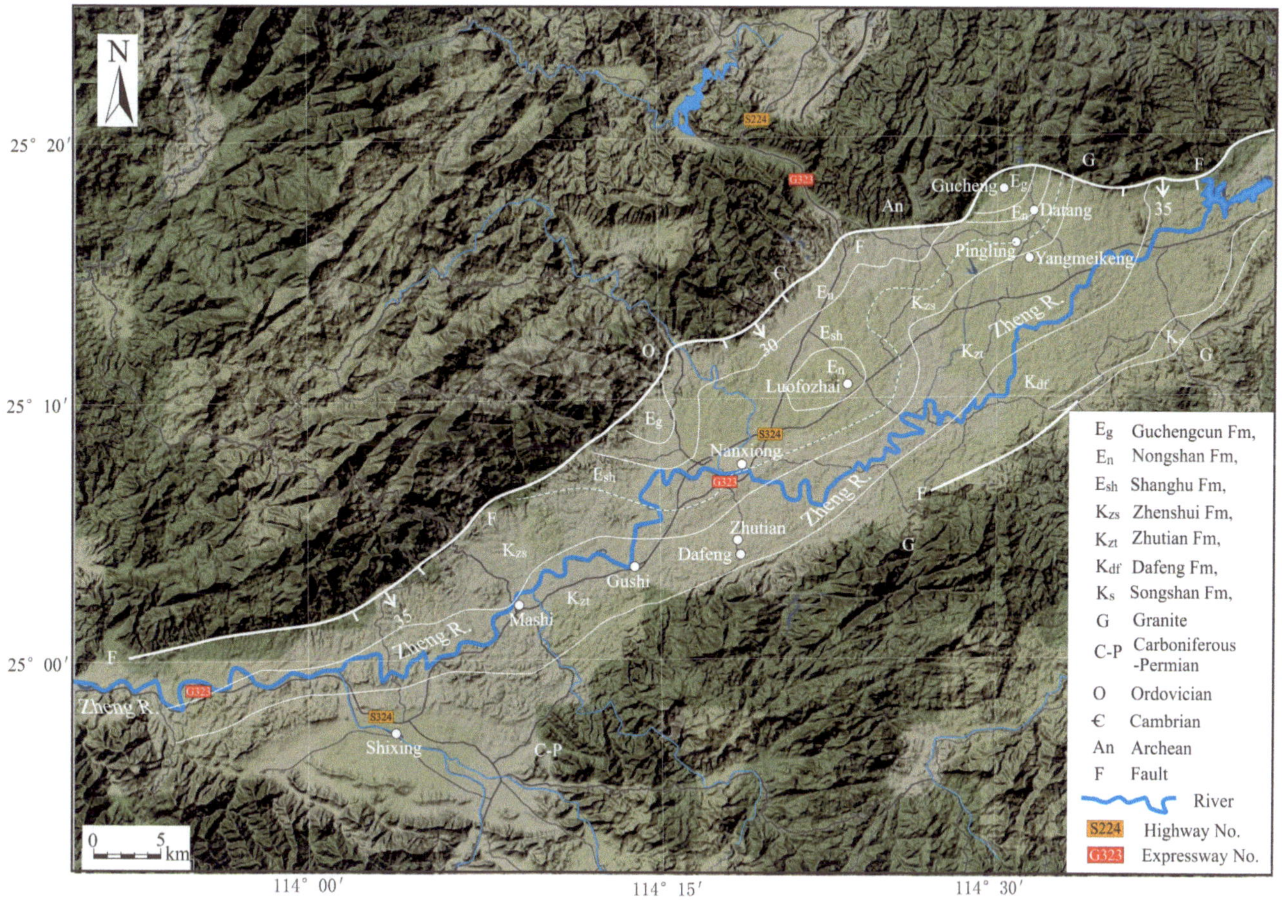

Fig. 7 Simplified geological and topographical map of the Nanxiong Basin, a fault control basin, i.e., a half-graben basin, adjacent to the 'Danxia basin' to the east

District is one of the typical spots that deserves detailed studies (Fig. 26).

Due to the availability of rich calcite cement, pseudo-karst micro-landforms can be easily studied in the red beds spots such as caves, arches and stalactite on cliffs and in caves (Huang et al. 1994; Fig. 27).

3.3 Evolution of Danxia Basin and Mount Danxiashan

Mount Danxiashan used to be an inland basin, named Danxia Basin in Cetaceous and survived until early Paleogene before dried up together with the Nanxiaong Basin. The basins then entered a long-term denudation period caused by the last phase of uplift of the Alpine tectonic cycle. Continuous exogenic erosion had transformed the ancient topography and basin into a peneplain by late Paleogene (Peng 2020a, Fig. 28a). The area was very much affected by the hard-collision (continental-continental collision) between Indian and Eurasian plates (Gibbons et al. 2015; Matthews et al. 2016; Suo et al. 2020); and by the northward movement of the eastward subduction under Eurasian of Pacific plate (Li et al. 2019). As a result, Neotectonic activities were triggered forcing the west part of Mount Danxiashan thrusted eastward along the Shaoguan-Renhua regional fault (F_1 in Fig. 26b). This led to rejuvenation of the land and the restart of the erosional cycle (Peng 2020a, infant stage in Fig. 28b). Since the late Pliocene, Mount Danxiashan area had entered into a predominantly exogenic process stage with the formation of mature landform due to gradual erosion (Peng 2020a, mature stage in Fig. 28c). During the period of tectonic uplift, some reaches of the Jinjiang River had incised along the Shaoguan-Renhua Fault (F_1) and others had cut down following their original courses and resulted in incised meanders.

4 Global Comparison

Red beds and their associated landforms are reported all over the world except Antarctica. In recent years, Danxia landscape has become extensively researched as a distinctive geomorphology. In terms of earth evolution, tectoni-

Fig. 8 Vertical cliff of Jinshiyan Temple, Mount Danxiashan

Fig. 9 Typical mesa and stone wall landscape, Zhanglaofeng scenery area, Mount Danxiashan

Fig. 10 Typical mesa and stone wall landscape, Yangyuanshan Scenic Area, Mount Danxiashan

Fig. 11 Camel Peak, a cuesta of Mount Danxiashan

Fig. 12 Pyramidal shape resembles a grand monument in Mount Danxiashan

Fig. 13 Round-top peak, Mount Danxiashan

Fig. 14 Stone peaks and cluster of rock columns around and the top of Teapot Peak in Baizhai Scenic Area, Mount Danxiashan

Fig. 15 Gigantic red bed column of Yangyuanshi (Male Rock), Mount Danxiashan

Fig. 16 Jinjiang River and red cliff at sunset in Mount Danxiashan

Fig. 17 Cluster of peaks and rock column, Mount Danxiashan

cally stable and mature continents were formed since the late Proterozoic, which provided the favorable environment for the formation of continental red beds. Therefore, the earliest red beds were discovered in the platform depression of late Proterozoic, such as the Pre-Cambrian red beds in the Australian and West African plateaus. Their formation was related to the drifting of early Gondwanaland to low latitude. After the Caledonian orogeny, the old, continental red sandstone (mainly molasse deposition) started to form in the sedimental basins in the mountains of North America, North Europe, Russia and South China (Zhang et al. 2011; Fig. 29). Red detrital rocks such as shales, sandstone and conglomerate were formed along the flanks of Appalachian geosynclines.

In late Paleozoic of Hercynian orogenic movement, the supercontinent, Pangea, was formed. Red beds and evaporate rocks were developed in many areas in the ancient continents of North America and Russia (Zhang et al. 2011; Fig. 29). The break-up of Pangea led to the collision of the North and South China continental plates. As a result, they were united in late Triassic to form a broad continental plate in East Asia, by the Indosinian orogenic movement. Since then, the continental red bed basins associated with this tectonic movement were formed, including those located in northern part of the South China plate.

Alpine orogenic movement is the equivalent of Yanshan orogenic movement in China. It was a very active global tectonic movement with the ancient Pacific Plates subducted

Fig. 18 A typical ravine (or called 'One-line sky' by Chinese) during sunset, formed along a vertical joint

Fig. 19 The most famous natural arch called Tongtai Bridge in Yangyuanshan Scenic Area, Mount Danxiashan

Fig. 20 Stratiform cave in Yanyuanshan Scenery Area. This flattened or horizontal cave was formed by differential weathering due to relative weak rock of siltstone, mudstone interbedded between relative strong rock layers of coarse sandstone or conglomerate

Fig. 21 These big cave clusters are apparently controlled by rock strata and shaped by differential weathering. They can be found in the north piedmont of Yangzhouzhai Peak, Mount Danxiashan

Fig. 22 Various Danxia peak assembly in Mount Danxiashan

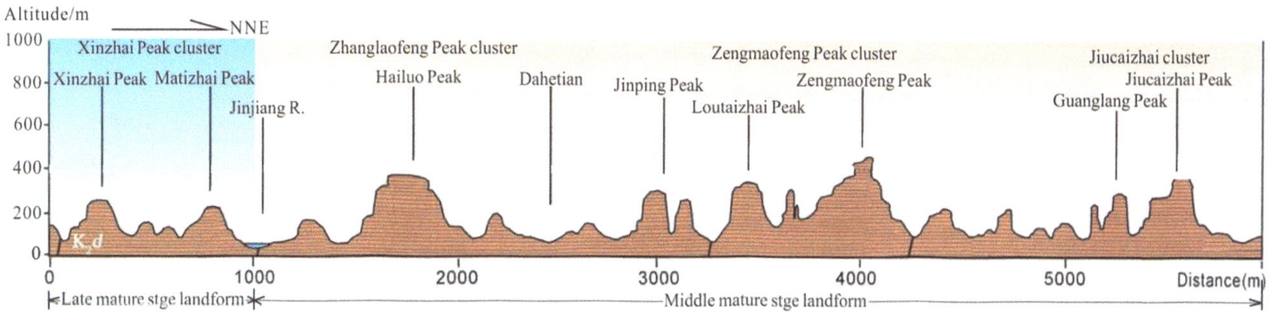

Fig. 23 Relative ages of peak-forests clustered in Mount Danxiashan. Late mature stage (left) and middle mature stage (right)

under the Eurasian Plate in the east, i.e., Izanagi Plate, as well as subducted into the American Plate in the west, i.e., Farallon Plate (Li et al. 2019). During this period, continental red basins distributed both in Eurasian Plate and North American Plate, especially in the North American, were mainly composed of fluvial and Aeolian sandstones. They formed the major Danxia landscape in the western part of North America (Figs. 30, 31). In China, the continental red beds were mainly composed of fluvial and littoral siltstone and mudstone which were hard to form Danxia landforms due to the weak resistant, fine detrital composition. However, during the Cretaceous, back arc area of North American and the western part of Europe began to tectonically subside, resulting in large scale northward transgression from the Gulf of Mexico in the south. It brought along with abundant marine chalk deposits during the Laramide Orogeny with relatively strong tectonic movement. In contrast, the subduction rates had slowed down in the eastern Pacific margin, with the extension regime started in eastern China. Large amount of inland red detrital rifted and fault basins had been formed under the tropical and subtropical environments in a striking NE pattern. They provided red bed materials to form the Danxia type landforms in the later time of Cenozoic. On the other hand, North Africa had started extension with the closing of Paleo-Tethys and opening of Neo-Tethys. The back arc area had developed red bed basins in Cretaceous, resulting in red sandstone landforms formed along the zone. The three places used to be in the tropics (Fig. 30).

Fig. 24 Landforms of different shapes and scales but having the same height, whether they are hills, towers or pinnacles, implying a relic planation surface (horizontal red dotted line), in Zhaolaofeng Peak Scenic Area, Mount Danxiashan

Fig. 25 Planation surface at 300 m in eastern Mount Danxiashan

Near east–west direction, the red bed basins were developed from North Africa, through middle Asia, Afghanistan for example, to East Asia including China along the neotethys and the westerly circulation mostly in Paleogene. They also appear on the east side of the Andes Range, induced by the eastern Pacific subduction. Therefore, Danxia landscape could be investigated in these two distinctive dry, climatic regions.

In summary, although the red beds could be formed in any times since late Proterozoic, the major red beds that could form typical Danxia landscapes concentrated on Cretaceous to Paleogene (Zhang et al. 2011; Fig. 32). The

Fig. 26 Tafoni developed along the fine detrital and siltstone layers between coarse detrital deposition of sandstone or conglomerate of Danxia formation, Jinghiyan Temple of Mount Danxiashan

Fig. 27 Stalactite in cave, Shanlikeng, Mount Danxiashan (Photographed by Huang Jin and sketched by Huang 1994)

preferred paleo-environments were terrestrial with hot and arid climate in areas of lower latitude, particularly with post orogenic basins of extensional regime and moderate uplift during the neotectonic periods. In other situations, these red beds can form the common 'red beds landform' but not the typical Danxia landforms due to relatively strong rock resistance or well compacted with more mature diagenesis and well silicic cements. As for the climates during the neotectonic or post diagenesis periods, it could either be continuously dry and hot such as the Danxia landscapes found in North America or have changed to become humid and warm such as those found in Southeast China.

5 Research History

5.1 History

As early as 1870s, during his field study from Guangzhou to Beijing, German scholar Ferdinand Paul Wilhelm Richthofen (1833–1905) had visited northern Guangdong. He recorded the red beds he had studied. Because the red beds overlied unconformably on all the older rocks and strata, he called them 'deck sandstone', which was probably the first academic name for Danxia red beds. He considered the red beds were early Tertiary in age based on the plant fossil, *Rhus atavia Shenk*,

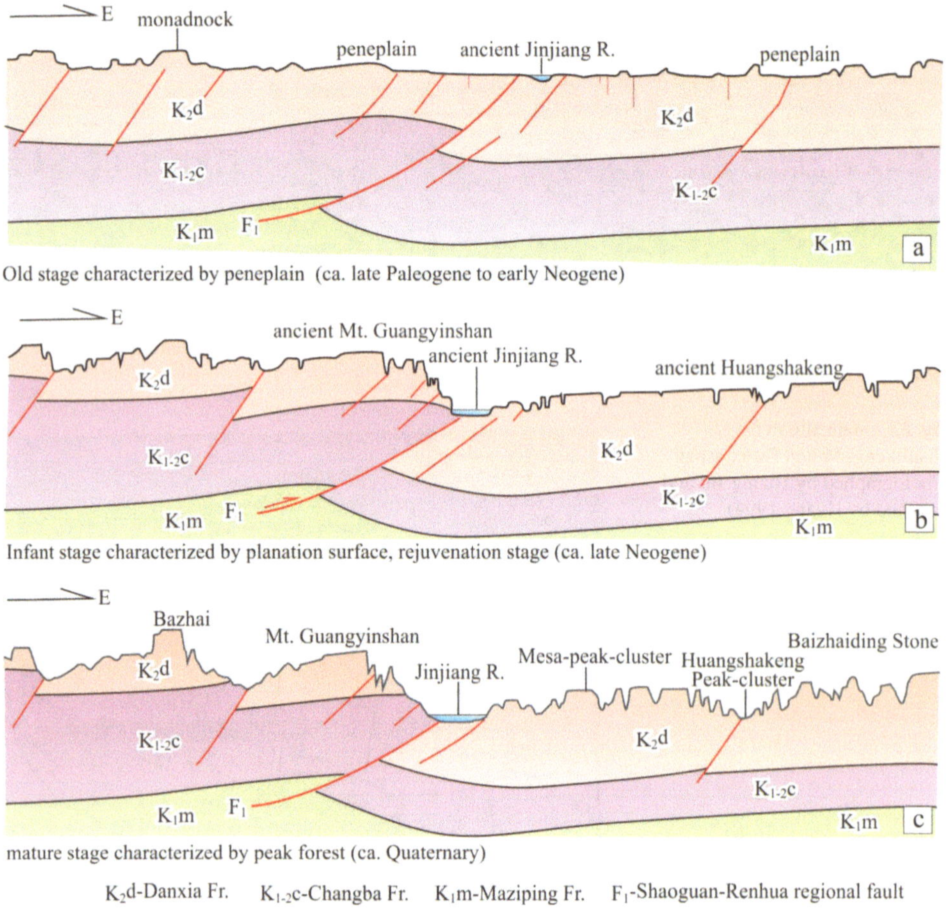

Fig. 28 Landform evolution stages of Mount Danxiashan (modified after Peng 2020a)

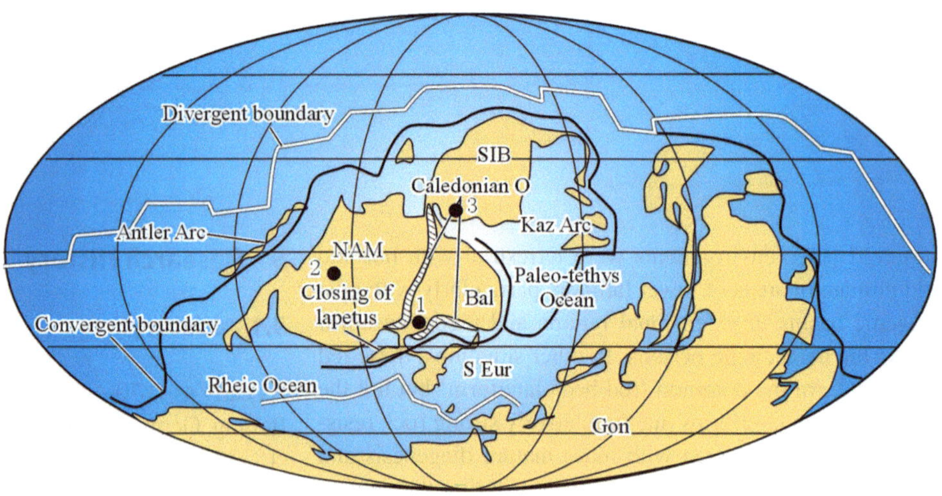

Fig. 29 Location of red beds in Devonian time. *Note* 1-Southwest Wales, 2-Western North America, 3- Canada Bal-Balranald, Gon-Gondwanaland, NAM-North America, SIB-Siberia

and had similar stratigraphy to that found in Sanshui, downstream of the North River (Chan 1938). Literally in Chinese, 'Danxia' is an image of 'reddish rays' or 'rosy clouds' appear during sunrise or sunset (Yan et al. 2019). The term 'Danxia' was initially designated by Chinese geologists in 1920s to classify specific stratigraphic sequence of red beds that form the sandstone hoodoos of the famous Mount Danxiashan in southern China (Fong and Zhu 1928; Yan et al. 2019). The name also reflected and described the natural beauty of the landscape.

With nearly 100 years of study and popularization of this specific landscape and acknowledging its aesthetic value, 'Danxia' has become a common Chinese name and the landforms associated with it have gradually become a unique landscape type similar to karst landscape (Peng 2020a, b). Danxia landscape can be characterized by red beds with flat top, steep

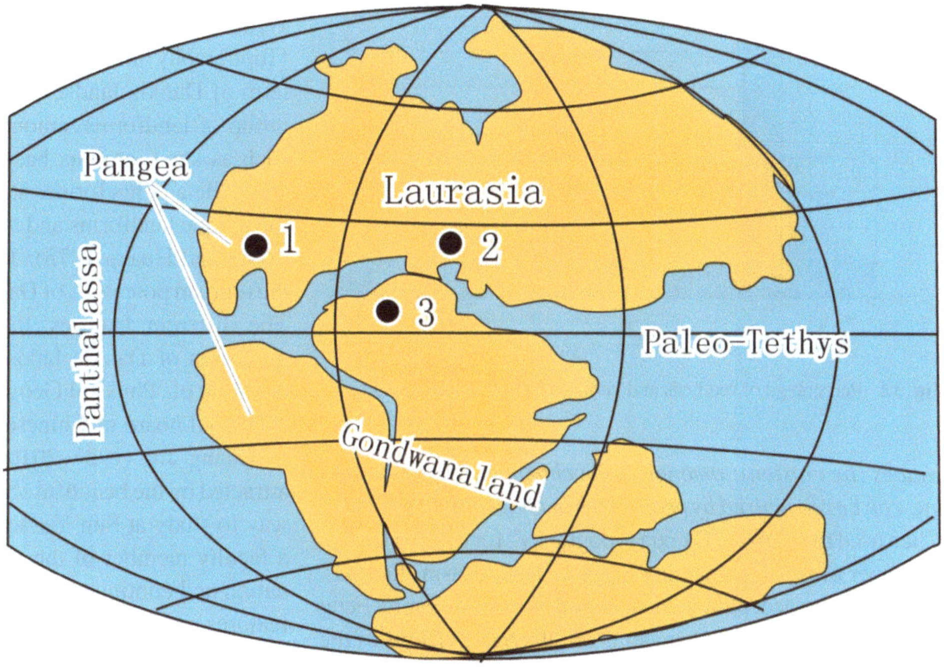

Fig. 30 Location of red beds in supercontinent, Pangaea in the Permian to early Triassic. 1-Western North America, 2-Western England, 3-Northwestern Africa

Fig. 31 Arches National Park in Utah, USA, is a typical Danxia landform in arid area. Entrada Sandstone is responsible for the formation of most of the arches and potholes

cliff and gentle foothill of the thick red beds formed under the conditions of neotectonic uplift and exogenic erosion. In 1928, Feng Jinlan (Fung, Kin-Lan, 1889–1976, Fig. 33) and Zhu Huisheng for the first time had named the red beds and related stratum in North Guangdong as 'Danxia formation' and made a vivid description of its lithology and landscape.

In 1933, Chen Guoda (Chan, Kuo-Ta, 1912–2004, Fig. 33) studied red beds in his bachelor thesis of 'The Red Beds of Kwangtung (Guangdong)'. His work was awarded by the National Institute of Geology of Academy Beiping (Beijing) in 1933 and published later by the Bulletin of National Institute of Geology of Academy Peiping in 1935 (Chan 1935). In 1938, Chen had described typical topography of Danxia formation (Tanhsia formation) in his paper of '*on the Subdivisions of the Red Beds of South-Eastern China*' as '*...sometimes fashioned the rocks into pillars and blocks of fantastic*

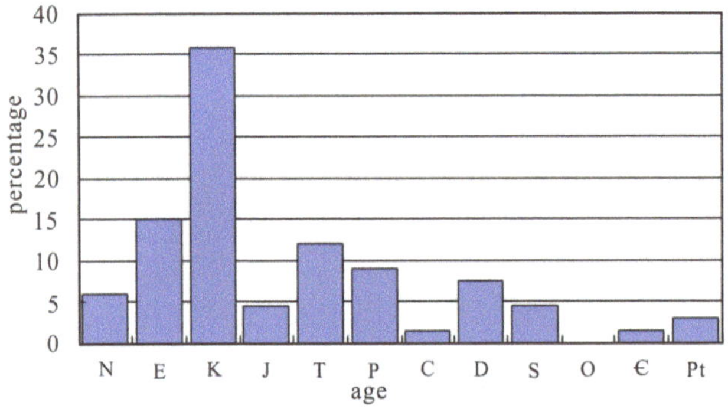

Fig. 32 Percentage of red beds and geological time in the world

shapes (by erosion); amongst them chimneys, giants' heads, etc. *can be recognized by people rich in imagination'*. In 1939, Chen used the term of 'Danxia topography" for the first time in his paper of the 'Geology in Gongshui River Basin, Jiangxi'. This term was accepted immediately in the same year by Feng Jinlan in his paper of 'Comments on 'the Subdivisions of the Red Beds of South-Eastern China' despite his disagreement with Chen on the proposed subdivisions (Feng 1939).

Wu Shangshi (Wu Shang-Shih, 1904–1947, Fig. 33) and Zeng Zhaoxuan (Tseng Chiu-Suen, 1921–2007, Fig. 33) were also pioneers of Danxia landscape study. During the invasion of Japanese army in 1940s, they moved to mountain area of northern Guangdong with Sun Yat-sen University to continue their education and field studies. Zeng Zhaoxuan used to be Wu's student and had field investigation with Wu most of time. Zeng had his bachelor thesis research on Mount Danxiashan in Renhua County of North Guangdong. Both had written large number of papers discussing the morphological features including Danxia landscape in northern Guangdong despite the tough living and research conditions during that war torn periods. Zeng had worked as a faculty member after his graduation in university in 1943. After World War II, they moved back to Guangzhou with the university and continued their education and research, but Wu unfortunately passed away in 1947 at the age of 43 due to overwork. But their works had continued to be published by Zeng after the death of Wu (e.g., Wu and Tseng 1946a, b, c, 1948, 1949; Tseng, i.e. Zeng, in Wade-Giles system, 1945). Zeng later published a book of 'Topography in China' in 1960 (Zeng 1960), in which the term of Danxia landscape was used to describe as a special group of landforms, among other types of landform groups such as karst, granite, basalt, sandstone, shale, laterites and loess. He and his former student, Huang Shaomin, discussed the unique landforms and their formation in a journal in 1978 (Zeng and Huang 1978). They also paid special attention to the tourism potentials of Danxia landscape in southeast China. The different features, distribution, petrology, developing processes of Danxia landscape had also been published in the book of 'Physical Geography of China: Geomorphology' (Editorial board of Chinese Physical Geography 1980).

Huang Jin (1927–2016, Fig. 33) had been very much attracted by the beautiful scenery of Mount Danxiashan on his way to study at Sun Yat-sen University in 1947. He became a faculty member of the Geography Department after graduation and continued his research particularly in quantifying features and processes associated with the formation of landforms of Danxia landscape. He took Danxia terrain as an independent geomorphological type in the Guangdong Geomorphologic Map and defined Danxia landscape for the first time during his talk in the First National Symposium of Tectonic Landform in 1981. His paper entitled 'Slope formation of Danxia landscape' was later published in 1984 (Huang 1984). His presentation summarized and characterized the landform features of Danxia landscape by highlighting its distinctive characteristics of flat top, steep cliff, gentle foothill which attracted nationwide attention and many following research in China after the Symposium. Huang's research had been supported by the Natural Science Foundation of China in 1990. Then in 1991, the First National Symposium on Danxia Landscape Tourism Development was held in Mount Danxiashan of Guangdong, and Danxia Geomorphology and Tourism Development Research Society of China had been set up. Thanks to the continuous efforts of Huang Jin and Peng Hua, research on Danxia landscape become very common in the country. Mount Danxiashan had successfully became a national geopark in 2002 and then became one of the first global geoparks in the country in 2004.

Fig. 33 Pioneers of Danxia landscape research (left to right): Feng Jinlan, Chen Guoda, Wu Shangshi, Zeng Zhaoxuan, Huang Jin and Peng Hua. They were all professors of Sun Yat-sen University, Guangdong

Fig. 34 China Danxia World Heritage Awards Ceremony in 2011

Peng Hua (1956–2018, Fig. 33) was a reputable scholar who popularized the concept of 'Danxia landscape' to the world and popularized it to become an academic term in the international geomorphological field. His passion began in 1987 during his first visit to Mount Danxiashan and became his lifetime research on this fascinating landscape. In May 2009, the first international academic conference on Danxia landscape was held in Mount Danxiashan and a 'Declaration of Danxia landscape' was announced. The conference affirmed the contribution of Chinese scholars, called for the recognition of Danxia landscape as an independent landscape type and encouraged international comparative studies. From then on, Danxia landscape has caught the attention of the international academic community. On July 7, 2009, during the 7th International Conference on Geomorphology in Melbourne, Australia, the IAG Council approved the establishment of the Danxia Geomorphology Working Group, signifying the recognition of Danxia landscape and its research. The objectives of the working group were to discuss and clarify the definition of such landscape type, as well as its formation, development, distribution and classification.

In December 2006, with support from the former national Ministry of Housing and Urban–Rural Development and the Provincial Construction Department of Hunan Province, the works of World Natural Heritage nomination started. Four years later, Danxia landscape has been successfully listed as a World Heritage during the 34th convention of the World Heritage on August 1st, 2010. Six typical and famous sites in South China had been successfully inscribed successfully in the List of World Heritage Sites (Fig. 34). Subsequently, research on Danxia landscape has increased significantly in China. Two international conferences were organized in China in 2010. The 2nd International Academic Symposium of Danxia Landscape was held in Mount Danxiashan in 2012, followed by the 3rd Conference of Sandstone Landscapes in Europe in 2013. Between 2013 and 2014, joint scientific research projects were conducted between China and Slovakia which offered joint degree for university students in both countries. In the eighth International Conference on Geomorphology in Paris in 2013, IAG had changed the working group name of 'Danxia Geomorphology' into 'Red Beds and Danxia Geomorphology' by adopting suggestion from Chinese scholars to extend research disciplines covering not only geomorphology but also geology, ecology, environmental sciences, geo-hazards, and associated land protection and conservation.

5.2 Potential Problems

Since the coining of the term of Danxia landscape (or Danxia topography), there has been no consensus about the definition among scholars. Since then, more than 20 definitions had been proposed (Wang et al. 2020). For the internationalization of the Danxia landscape concept and to encourage global comparative studies, it is important to clarify the definition of Danxia landscape. More research on the different mechanisms, particularly in microscale on mineralogy, geochemistry, diagenesis, biology (e.g., rock surface moss) for reddish color origin and its range, should be emphasized rather than just on local tourism development. Lastly, much stronger efforts should be put in promoting Danxia landscape instead of concentrating only in China.

5.3 Recent Research and Future Prospective

The research on Danxia landscape is continuous, both qualitative and quantitative assessments on the geomorphic characteristics. Quantitative analyses comprise digital geomorphology or DEM models (e.g., Zhang et al. 2018), rock resistant measurement (Zhu et al. 2010), experimental study (Zhu et al. 2015) and TCN dated erosional landform ages. Other research include macro to micro scales forming mechanism of cave or tafone in Danxiashan (Chen et al. 2019), the iconic reddish color with color range measured by spectrograph using Mount Danxiashan as a standard and the origins of color mineral (e.g., Luo et al. 2021; Yang et al. 2022; He et al. 2022). In addition, studies are also conducted on the changes of depositional, diagenetic and post-diagenetic environments, biological factors such as ecology and global warming on color and landform changes.

Prioritized research is also on developing a discriminable definition system used to assess and identify landforms which are to be classified as Danxia to Danxia-like landforms, using Mount Danxiashan as a standard. Techniques such as spectroscopies of mineral such as microcosmic or mineralogical techniques such as XRD, EPMA, LA-ICP-MS, XPS, FTIR, HRTEM, Raman and photic spectrometer or macroscopic and remote sensing techniques such as ArcGIS, LiDAR, High spectrometer. Cross-discipline study is also encouraged to broaden the research of Danxia landscape.

6 Meeting the Criteria of UNESCO World Heritage Convention and Global Geopark

6.1 World Heritage Convention

Mount Danxiashan meets the World Heritage Convention's criteria (Dingwall et al. 2005) in (vii), (viii), (ix) and (x) (Peng 2020a):

Criterion (vii). Scenically superlative landscape and landforms, displaying natural environment elements, such as huge red rocks of vertical walls, cliffs, peaks, towers, mesas, cuestas, pillars, caves, arches, gorges, ravines, as well as forests and water, with exceptional natural esthetic value under warm and humid monsoon climate.

Criterion (viii): Vitally important step in the process of evolution of the earth, including the geological origin, the ongoing distinctive landform evolution, and geomorphic processes, and biological lives on the landforms, recording tectonic processes from terrestrial depressive environment in late Mesozoic to tectonic uplift in Cenozoic coeval with climate changes.

Criterion (ix): Different topographies harmonizing with various flora-fauna ecosystem provided by isolated or fragmented hill tops, cliffs, or ravines, in subtropical forest and rainforest, especially ecosystem for refuges of rare, endemic or unique, endangered and species with some being called 'living fossil', such as *Firmiana Danxiaensis*, *Primulina huangjiniana* and *Diospyros penghuae* (Hsue et al. 1987; Huang et al. 2020; Zhao et al. 2018).

Criterion (x): Conservation *in-situ* of the highest biodiversity and eco-diversity in so called 'Indo-Malayan' and 'Southeast China-Hainan Subtropical Moist Forest'.

6.2 Global Geopark

Mount Danxiashan meets the following criteria of Global Geoparks Network:

Criterion (i): Not only of geological significance but also of ecological, archaeological, historical and cultural values.

Criterion (ii): Sustainable development on tourism, socio-economic and culture by local involvement or with local harmonic benefits, i.e., UNESCO, local authorities, local population, national Geological Surveys, etc.

Criterion (iii): Stimulate economic activities with the framework of cultural and environmentally sustainable development interaction with geological heritage protection.

Criterion (iv): Geo-education using local information about geology, geomorphology, physical geography and heritage components.

Criterion (v): On one hand, protection of geological heritage in accordance with local traditions and legislative obligations; on the other hand, conservation of geological features such as representative rocks exposures, minerals, fossils and landforms, in accordance with national legislation of regulations, providing information on various geoscientific disciplines.

Criterion (vi): As a member of Global Geoparks Network (GGN), a geopark gains worldwide recognition and benefit through the exchange of knowledge and expertise, experience and staff among geopark network, and still shares common values, interests, or backgrounds, to develop a specific methodology and management practices (UNESCO 2010).

7 Conclusions

Danxiashan Geopark is a nominated place of Danxia landscape, which is a new type of sandstone landscape and landform category very popular in China and increasingly being noticed around the world. It is prominently recognized as representative scenery of beautiful sunset and sunrise color. This typical landscape carried various stunning and attractive landforms such as steep cliffs, castle-like peaks, valleys and caves, providing a special ecological environment for abundant endemic, rare, imminent species of flora and fauna, forming unique ecosystem under subtropics humid and warm climate. This distinctive landscape has also been used as an ideal venue for religious activities, tourism, and refuge during war-torn periods. Therefore, Danxiashan Geopark not only has great values in geology and geomorphology, but also in ecology, archaeology, history and culture. It not only preserves these natural and human heritages and at the same time fosters local economic development, environmental conservation, geoscience education and international exchange. The meeting of the criteria of both World Heritage Convention and Global Geoparks Network has proven that Mount Danxiashan, the Redstone Park of China, is an important asset of its people as well as of the world.

References

Allison RJ (1993) The sensitivity of landscape. John Wiley & Sons Ltd., New York

Chan K (1935) The red beds of Kwangtung. Bull Inst Geol Acad Peiping 6(4)

Chan K (1938) On the subdivisions of the red beds of South-Eastern China. Bull Geol Soc China 18(3–4):301–322

Chen G, Liu H (1939) Geolgoy of Gongshui Drainage Basin, Jiangxi Province. Jiangxi Geological Collection 2 (in Chinese)

Chen L, Guo F, Liu F et al (2019) Origin of Tafoni in the Late Cretaceous Aeolian Sandstones, Danxiashan UNESCO Global Geopark, South China. Acta Geol Sin (English Edition) 63(2):451–463

Davis WM (1909) Geomorphic essays (Johnson DW (ed). Boston Ginn & Co, Boston (Report 1954)

DCCEEW (2025) Geology of Uluru and Kata Tjuta. https://www.dcceew.gov.au/parks-heritage/national-parks/uluru-kata-tjuta-national-park/natural-environment/geology. Accessed 12 June 2025

Dingwall P, Weighell T, Badman T (2005) A contribution to the global theme study of world heritage natural sites. IUCN

Editorial board of Chinese physical geography (1980) Physical Geography of China: Geomorphology. Science Press, Beijing (in Chinese)

Feng J (1939) Comment on "the Subdivisions of the Red Beds of South-Eastern China." Geol Rev 4(1):173–184 (in Chinese)

Fong KL, Zhu HS (1928) Preliminary report on the geology and mineral resources of Chu Chiang, Jen Hua, Nan Hsiung and Shih Hsing Districts of Northern Kwnagtung. Geol Surv Kwangtong Kwangsi 1:29–54

Geologyin (2025) The rainbow mountains in Peru, Vinicunca. https://www.geologyin.com/2016/02/the-rainbow-mountains-in-peru-very.html. Accessed 20 June 2025

Gibbons AD, Zahirovic S, Müller RD et al (2015) Tectonic model reconciling evidence for the collisions between India, Eurasia and intra-oceanic arcs of the central-eastern Tethys. Gondwana Res 28(2):451–492

Guo F, Chen L, Yan Z et al (2020) Definition, classification, and Danxianization of Danxia landscapes. Acta Geol Sin 94(2):361–374 (in Chinese with English abstract)

He W, Yang Z, Du H et al (2022). Micro-mechanisms and implications of continental red beds. Minerals 12:1–14

Hsue H, Hao K, Hua S, Xu S (1987) New species and variety of Sterculiaceae from China. J South China Agric Univ 8(3):1–5 (in Chinese with English abstract)

Huang C, Meng K, Guo J et al (2020) Primulina huangjiniana (Gesneriaceae), a new species from Guangdong, China. Guihaia 40(10):1429–1437

Huang J (1984) A basic-forming way for Danxia landform slope. In: Geomorphology Committee of Geographical Society of China (ed) A selected collection of the first national symposium on tectonic landform. Science Press, Beijing, 193–195

Huang J (2010) Landforms in Mt. Danxiashan. Science Press, Beijing (in Chinese)

Huang J, Zhu S, Liang B (1994) Preliminary study on Pseudo-karst landforms in Mt. Danxiashan. Econ Geogr 14 (suppl.):27–32 (in Chinese)

IUGS (2025) The Jurassic Navajo sandstone at Coyote Buttes and the wave. https://iugs-geoheritage.org/geoheritage_sites/the-jurassic-navajo-sandstone-at-coyote-buttes-and-the-wave/. Accessed 10 June 2025

Li S, Cao X, Wang G et al (2019) Meso-Cenozoic tectonic evolution and plate reconstruction of the Pacific Plate. J Geomech 25(5):624–667 (in Chinese with English abstract)

Luo X, Yang Z, Zhang K et al (2021) A mineralogical study on the genesis of red color of the Danxia Mountain in Guangdong Province. Acta Mineral Sin 41(6):704–712 (in Chinese with English abstract)

Matthews KJ, Mullner RD, Sandwell DT (2016) Oceanic microplate formation records the onset of India-Eurasia collision. Earth Planet Sci Lett 433:204–214

Peng H (2000) A survey of the Danxia landform research in China. Sci Geogr Sin 20(3):203–211 (in Chinese with English abstracts)

Peng H (2020a) China Danxia. Higher Education Press, Beijing (in Chinese)

Peng H (2020b) Danxia geomorphology. Science Press, Beijing, Beijing

Phillips FD (2007) The perfect landscape. Geomorphology 84:159–169

Suo Y, Li S, Cao X et al (2020) Two-stage eastward diachronous model of India-Eurasia collision: constraints from the intraplate tectonic records in Northeast Indian Ocean. Gondwana Res 102:372–384

UNESCO (2010) Guidelines and criteria for national geoparks seeking UNESCO's assistance to join the Global Geoparks Network, UNESCO

UNESCO (2022) Operational guidelines for the implementation of the world heritage convention. United Nations Educational, Scientific and Cultural Organization (UNESCO), Inter-Governmental Committee for the Protection of the World Cultural and Natural Heritage, World Heritage Centre

USGS (2025) Geology of Bryce Canyon National Park, Geology & Ecology of National Parks. https://www.usgs.gov/geology-and-ecology-of-national-parks/geology-bryce-canyon-national-park#:~:text=The%20Tertiary%20rocks%20are%20primarily,well%20as%20sandstone%20and%20shale. Accessed 20 June 2025

Wang R, Ji S, Lin J (2020) On the definition of Danxia landform. China Terminol 22(3):60–65 (in Chinese with English abstract)

Wu SS, Tseng CS (1946a) The red beds in North Kwangtung. Lingnan J (Special issue for Wu Shang-shih and Tseng Cui-Suen in English) 12–20

Wu SS, Tseng CS (1946b) A geomorphology of Yuenhua Hsien. Lingnan J (Special issue for Wu Shang-shih and Tseng Cui-Suen in English) 1–11

Wu SS, Tseng CS (1946c) Preliminary observations on the geomorphology of Tung Kiang. Lingnan J (Special issue for Wu Shang-shih and Tseng Chui-Suen in English) 21–26

Wu SS, Tseng CS (1948) On the principal erosion surfaces and transversal georges of North Kwangtung. Lingnan J 9(1)

Wu SS, Tseng CS (1949) The Red Basin of Pingshek, North Kwangtung: A physiographic study. Lingnan J 9(1)

Yan L, Peng H, Zhang S et al (2019) The spatial patterns of red beds and Danxia landform: implication for the formation factors-China. Sci Rep 1961(9):1–10

Yang Z, Du H, Luo X et al (2022) Mineralogy study on the origin of white spots in the red strata of Danxia Mountain, Guangdong. Miner Resour Geol 36(1):129–137 (in Chinese with English abstract)

Tseng CS (1945) Preliminary research on the landform in red thick sandstone, South Renhua. Geogr Coll 12(in Chinese)

Zeng Z (1960) Petrotopography. Geological Publishing House, Beijing (in Chinese)

Zeng Z, Huang S (1978) The red bed landform in Southeast China. J South China Norm Univ (Nat Sci Ed) 1978(1):57–73; 1978(2):56–73

Zhang G, Chen K, Zhang H et al (2018) The evolution stage decision of Danxia landform based on digital elevation model (DEM). Acta Sci Natur Univ Sunyatseni 57(2):12–21 (in Chinese with English abstract)

Zhang K, Huang Y (1995) Research on the Planation Surfaces in North Guangdong. Trop Geogr 15(4):296–305

Zhang K, Yu Z, Hu Q et al (2011) The red beds and their landforms around the world. In: Collected papers of the second international symposium on Danxia landform and of the first conference of Danxia Geomorphology Working Group of IAG, pp 50–56

Zhao W, Fan Q, Chen F et al (2018) Diospyros penghuae, a new species of Ebenaceae from Danxiashan mountain in Guangdong, China. Acta Sci Nat Univ Sun Yat-Sen 57(8):98–103

Zhu C, Peng H, Ouyang J et al (2010) Rock resistance and the development of horizontal grooves on Danxia slopes. Geomorphology 123(2010):84–96

Zhu C, Wu L, Zhu T et al (2015) Experimental studies on the Danxia landscape morphogenesis in Mt. Danxiashan, South China. J Geogr Sci 25(8):943–966

Ecology

Wenbo Liao and Ting Zhou

Abstract

Spanning an area of 292 km^2, Mount Danxiashan lies at the northern edge of the South Subtropical Zone and the southern foothills of the Nanling Mountains. It is characterized by over 680 red sandstone peaks, notable for their flat tops, steep cliffs, and gentle bases. The mountain's striking red hues, complemented by lush vegetation at both its base and summit, create a stunning landscape featuring a harmonious blend of clear waters, green trees, and vibrant red rocks. Situated in the tropical monsoon climate region of South Asia, Mount Danxiashan enjoys mild winters and summers free of extreme heat. As one of China's rainy regions, it records an average annual precipitation of 1,715 mm.

1 The Unique Ecological Environment of Mount Danxiashan

Spanning an area of 292 km^2, Mount Danxiashan lies at the northern edge of the South Subtropical Zone and the southern foothills of the Nanling Mountains. It is characterized by over 680 red sandstone peaks, notable for their flat tops, steep cliffs, and gentle bases. The mountain's striking red hues, complemented by lush vegetation at both its base and summit, create a stunning landscape featuring a harmonious blend of clear waters, green trees, and vibrant red rocks. Situated in the tropical monsoon climate region of South Asia, Mount Danxiashan enjoys mild winters and summers free of extreme heat. As one of China's rainy regions, it records an average annual precipitation of 1,715 mm.

The mountain is traversed by the Jinjiang and Zhenjiang River basins, flowing from north to south. The Jinjiang River winds through the Danxia landform area, while the Zhenjiang River encircles the mountain in the southeast. The two rivers converge at Wu Ma Ying Bin, located in the southern part of Mount Danxiashan, forming the upper reaches of the North River. From there, they flow southward, merging with the Xijiang and Dongjiang Rivers in San Shui, Guangdong, before continuing into the lower Pearl River Basin. The Nanling Mountains serve as an ecological barrier in northern Guangdong. Warm air currents from the south and tropical-subtropical flora travel along the river valleys into the mountain's depths, fostering a monsoon forest ecosystem dominated by species such as *Castanopsis jucunda*—Fagaceae (Beech and Oak family), *Artocarpus hypargyreus*—Moraceae (Mulberry and Fig family), and *Ormosia semicastrata*—Fabaceae (Legume or Pea family).

This phenomenon is aptly named the 'Moist Danxia Valley Effect'.

2 Vegetation and Ecosystem

2.1 Vegetation Types

The vegetation landscape of Mount Danxiashan is a rich mosaic of multiple ecosystems. Based on the principles of phytocoenology and ecology, the vegetation is classified into nine types, 29 formations, and 47 associations. The primary vegetation types include I. Warm coniferous forest II. Warm deciduous broad-leaved forest III. South subtropical monsoon evergreens broad-leaved forest IV. Subtropical evergreen broad-leaved forest V. Montane sclerophyllous broad-leaved forest VI. Subtropical montane deciduous shrubs VII. Subtropical mountain shrubs and

W. Liao (✉)
School of Life Sciences, Sun Yat-Sen University, Guangzhou, Guangdong Province, China
e-mail: lsslwb@mail.sysu.edu.cn

T. Zhou
National Park and Nature Education Research Institute, Sun Yat-Sen University, Guangzhou, Guangdong Province, China

Y. Ng and W. Eder (eds.), *The Redstone Park of China*, Geoheritage, Geoparks and Geotourism,
https://doi.org/10.1007/978-3-031-95603-4_5

grasslands VIII. Wetland communities IX. Artificial communities (Zhou et al. 2019).

Horizontally, the most significant vegetation types are the warm coniferous forest, dominated by *Pinus massoniana*—Pinaceae (Pine family), and the subtropical monsoon evergreen broad-leaved forest, featuring *Castanopsis jucunda*—Fagaceae (Beech and Oak family) and *Schima superba*-Theaceae (Tea family). Vertically, the complex peaks intensify local thermal radiation, creating a 'hot valley' effect—dry and hot summers alongside wet and warm winters. Tropical elements ascend through the Pearl River basin and valleys to the mountain tops. Below 250 m, subtropical monsoon evergreen broad-leaved forests appear, occasionally forming layers of subtropical valley rainforest. Valleys and lowlands host warm deciduous broad-leaved forests, typified by *Salix mesnyi*-Salicaceae (Willow family) (Zheng et al. 2019).

Between 250 and 350 m, subtropical evergreen broad-leaved forests dominated by *Castanopsis eyrei* flourish. As elevation increases, thinner soil layers and more cliffs nurture sclerophyllous evergreen broad-leaved elfin forests and deciduous shrubs, such as *Quercus phillyreoides*—Fagaceae (Beech and Oak family) and *Lagerstroemia indica*—Lythraceae (Loosestrife family).

At higher altitudes (350–625 m), locations like Hailuofeng and Bazhai host arid montane evergreen alfin forests (*Vaccinium bracteatum, Rhodomyrtus tomentosa*) and shrublands characterized by rupestrine species, including *Selaginella* and *Boraginella.*

In peripheral areas, terraces, and low valleys, artificial forests (e.g., *Cunninghamia lanceolata*, bamboo, eucalyptus), orchards (lychee, waxberry, citrus), and agricultural landscapes (paddy fields) dominate.

Bamboo forests are a distinctive and continuous feature of Mount Danxiashan, comprising 45 species from nine genera. These include 25 species of *Bambusa*, seven of *Phyllostachys*, five of *Dendrocalamus*, and others such as *Pseudosasa*, *Acidosasa*, and *Indocalamus*. Species like *Phyllostachys reticulata*—Poaceae (Grass family) and *Bambusa multiplex*—Poaceae (Grass family) are widely distributed. In 2019, two new bamboo species, *Phyllostachys danxiaensis* and *Dendrocalamus pulverulentoides*, were identified, further enriching the region's botanical heritage (Zheng et al. 2019).

2.2 Ecosystem

2.2.1 Ecosystem Diversity

The ecosystem of Mount Danxiashan is remarkably diverse, comprising 42 distinct types. These include 11 natural ecosystems, such as forests, grasslands, and still-water systems; 14 artificial ecosystems, like artificial wetlands, forests, agricultural-natural composites; and 17 composite ecosystems, including artificial and semi-natural blends (Fan et al. 2015; Li 2018).

Among these, Mount Danxiashan features unique ecosystem types, such as the cave grassland, rocky grassland, and mountain spring ecosystems. The distinct formation process of the Danxia landforms has created caves of various shapes and sizes, offered ideal shelter and feeding grounds for birds and cave-dwelling animals. This has given rise to the Danxia cave grass ecosystem, a specialized habitat within the landscape.

Additionally, vine plants growing in thick soil layers at the base of Danxia slopes ascend steep rocks using suckers or adventitious roots. This process improves the harsh conditions of exposed cliffs, enabling mosses, ferns, and other small grasses and shrubs to establish themselves. Over time, this fosters primary and secondary ecological succession on the cliffs, resulting in another distinctive ecosystem of the Danxia landscape: the cliff grass ecosystem.

2.2.2 Ecological Effects of Danxia Landscape

The danxia landscape exhibits unique ecological effects tied to its ravines and hilltops, creating a highly diversified cliff system on a small scale that evolves into distinct ecosystems. Mountaintops are characterized by xerophytic vegetation, while mountainside cliffs predominantly support lichens, subaerial algae, and xerophytic lianas. In cliff depressions and caves, short-lived and moisture-loving herbaceous plants thrive, whereas the mountain valleys host moist seasonal rainforests, evergreen broad-leaved forests, and bamboo forests.

Through ongoing water erosion, the danxia landscape has developed numerous river valleys, such as the Jinjiang and Zhenjiang River basins, forming hundreds of lowlands. Warm air and subtropical and tropical elements flow up the river valleys into the heart of Mount Danxiashan, fostering monsoon forests dominated by *Castanopsis jucunda*, *Artocarpus hypargyreus*, *Ormosia semicastrata*, and *Saurauia tristyla*. This phenomenon, known as the 'moist danxia valley effect,' supports notable species like tree ferns (*Alsophila spinulosa*, *Alsophila podophylla*), primitive *Eusporangiopsida* (*Ophioglossum vulgatum*), Marattiaceae (*Angiopteris fokiensis*), and tropical woody vines (*Gnetum*, *Fissistigma*, *Uvaria*, *Mucuna*, *Millettia*, Bauhinia) (Fig. 1).

Ecological phenomena such as strangulation, stem flowering, and aerial epiphytism are common in species like *Ficus*, *Artocarpus*, and *Drynaria roosii*. Species such as *Endospermum*, *Pandanus*, and *Calamus* migrate into the Nanling Mountains along the North River Basin or through the Guangdong-Jiangxi Corridor. These communities, shaped by the ravine effect, form stable successions aligning with the landform's ecological conditions, emerging as top-level communities.

1. *Lysimachia danxiaensis* 2. *Bambusa pervariabiloides* 3. *Asplenium guodanum*
4. *Caridina danxiaensis* 5. *Sinamma danxia*
6. *Satorius cheni* sp.nov. 7. *Agraphydrus danxiaensis* sp. nov. 8. *Orthobula nigre* sp. Nov.

Fig. 1 Rare species of Mount Danxiashan

Temperate species like *Salix mesnyi* also extend into Danxia valleys via the Guangdong-Jiangxi Corridor and the North River Basin, forming distinctive *Salix mesnyi* communities. River valleys, lowlands, and swamp wetlands preserve key species such as *Nymphoides coreana*—Menyanthaceae (Floatingheart family), *Schizocapsa plantaginea*—Dioscoreaceae (Yam family), *Typha orientalis*—Typhaceae (Cattail family), and *Vallisneria natans*—Hydrocharitaceae (Frog's-bit family).

On high-altitude peaks, clusters, forests, and pillars of red-bed rock develop. Hilltops endure thermal radiation, sandy soils, steep slopes, and strong winds, creating an island effect that supports xerophytic species like *Osteomeles subrotunda*, *Lagerstroemia* indica, *Hemerocallis fulva*, *Selaginella tamariscina*, *Quercus phillyreoides*, and *Vaccinium bracteatum*. Certain species, such as *Radermachera sinica*, *Citrus mangshanensis*, and *Cycas fairylakea*, exist only in isolated peaks. Notably, *Cycas fairylakea*, discovered in the Danxia landscape, has a population of just 10 individuals (Qin et al. 2017).

Within the danxia landscape, mountaintops generally exhibit higher temperatures and lower humidity compared to valleys and foothills. Species diversity on mountaintops is also lower, with valley vegetation showing stronger tropical characteristics. These features distinguish Danxia landscapes from typical non-danxia mountains. Ecological studies reveal significant differences in species composition, leaf area, crown width, and other factors between mountaintop and foothill communities. The hilltop and ravine effects hold significant scientific value for studying plant life forms, ecological niches, island biogeography, and adaptive evolution.

2.2.3 Vegetations of Danxia Landscape

The vegetation landscape of Mount Danxiashan is a striking and unique natural ecological feature among regions with Danxia landforms. Its distinctive vegetation has developed over time, shaped by the interaction of landforms, geological parent rock, ecological conditions, and biological factors, ultimately creating harmonious plant communities that match their habitats. According to Professor Zhen Li, ten types of landscape vegetation have been identified, named following the standard format of (landform/habitat) + (vegetation type/ biological landscape). These are:

1. Green mountain tops: While the cliffs of Mount Danxiashan are steep and forest-free, the flat mountaintops are covered in lush green vegetation. Subtropical evergreen broad-leaved forests, montane monsoon evergreen broad-leaved forests, and sclerophyll elfin forests dominate here, featuring species such as Quercus *phillyreoides*, *Castanopsis eyrei*, *Schima superba*, *Vaccinium bracteatum*, and *Rhododendron simsii*.
2. Forest in the platforms: Evergreen broad-leaved forests are widespread on the mid-to-upper layers and gentle slopes of large mountains, with an array of tree species including *Castanopsis jucunda*, *Schima superba*, *Ormosia semicastrata*, *Altingia chinensis*, *Engelhardia roxburghiana*, *Pinus massoniana*, and *Cinnamomum camphora*.
3. Green forests in the hills: These forests provide the foundational greenery for Danxiashan's hills and mountains. Key communities include mixed forests of *Pinus massoniana* and *Schima superba*, along with artificial pine, cedar, and bamboo forests (Qin et al. 2017).
4. Rainforest in the valleys: This layer of subtropical monsoon evergreen, broad-leaved forest lines valleys and streams, showcasing species like *Castanopsis jucunda*, *Artocarpus hypargyreus*, *Calamus thysanolepis*, Angiopteris fokiensis, *Musa basjoo*, and *Reevesia thyrsoidea*.
5. Bamboo forests by rivers and streams: These bamboo groves, trees, and shrubs along waterways highlight the scenic Jinjiang River. Dominant species include *Bambusa pervariabilis*, *Bambusa textilis*, *Salix mesnyi*, and *Triadica sebifera* (Niu et al. 2020).
6. Shrubs on the cliffs: Low trees, shrubs, and grasses thrive in cliffs, rock niches, and exposed areas. Typical species include *Osteomeles subrotunda*, *Lagerstroemia indica*, and cryptogamous plants like lichens and mosses, as well as flowering communities like *Hemerocallis fulva*.
7. Grass and trees on sunny slopes: Shrubs and sparse forests dot hilly sunny slopes, with species like *Rhodomyrtus tomentosa*, *Loropetalum chinense*, and *Phyllostachys sulphurea*.
8. Grasses on terraces: Tall grasses and reeds inhabit terraces, riverbeds, and gravel beaches, with dominant species such as *Saccharum arundinaceum* and *Miscanthus floridulus*.
9. Ancient trees in the countryside: These long-living trees are iconic in villages and temples, such as the 700-year-old *Ficus microcarpa* in Shuinan Village or the 370-year-old *Cinnamomum camphora* at Shuinan Ancient Pagoda.
10. Paddy fields in the countryside: Agricultural landscapes of paddy fields, sugarcane, vegetable fields, and orchards feature prominently, with Xiafu Village offering a picturesque view framed by peaks and the Jinjiang River.

The impressive red rocks adorned with vibrant greenery, secluded low valleys, verdant shrubs on high cliffs, and rivers reflecting bamboo forests paint a captivating picture of the idyllic countryside of Mount Danxiashan.

2.2.4 Characteristics of Vegetation Succession

The danxia landscape exhibits complete primary and secondary succession series, offering a remarkable ecological setting. The peaks of this iconic terrain demonstrate the primary succession of elfin shrubberies and forests. On the surrounding slopes, ongoing seasonal rainfall erosion, weathering, and gravity-induced collapses initiate primary succession on exposed rocks. This process begins with the formation of bryophyte and herbaceous plant communities, which mark the early stages of succession. As rocks further weather and bryophytes, lichens, and other plants contribute to soil development, the succession advances. The unique geological and geomorphological processes lead to recurring rock collapses, restarting primary succession on newly exposed surfaces and creating varying time series of succession stages.

Additionally, the danxia landscape supports a full secondary succession series, encompassing pioneer forests, transition forests, and stable forest communities. This dual succession system provides a rare and ideal environment for ecologists to study an uninterrupted sequence of vegetation development within a single location. It offers valuable insights and guidance for the protection, management, sustainable use, and restoration of vegetation in the region.

3 Flora

3.1 Composition and Characters of Flora in Mount Danxiashan

Between 2017 and 2022, researchers from Sun Yat-sen University conducted an extensive investigation into the biodiversity of Mount Danxiashan. The study identified 2,048 species of higher plants (including mosses, ferns, and seed plants) across 915 genera in 244 families. Additionally, there are 181 cultivated plant species belonging to 143 genera and 63 families (Peng et al. 2011; Liao 1992). The flora's composition is summarized as follows:

3.1.1 Bryophytes

The study identified 161 species across 97 genera in 52 families, including 52 liverwort species, 1 hornwort species, and 108 moss species. Subtropical families dominate, such as Ditrichaceae, Dicranaceae, Leucobryaceae, and Thuidiaceae. Warmer, wetter valley lowlands support South Asian tropical families like Meteoraceae and Sematophyllaceae, while temperate families such as Pottiaceae are limited to exposed gravel habitats.

3.1.2 Pteridophytes

The researchers documented 154 species across 63 genera in 27 families, primarily tropical and subtropical families, including Polypodiaceae, Selaginellaceae, Adiantaceae, and Dryopteridaceae. These families are mostly distributed in South and Southwest China.

3.1.3 Seed Plants

The study revealed 1,733 seed plant species from 755 genera in 165 families, including 9 gymnosperms and 1,724 angiosperms. Representative families include Theaceae, Fagaceae, Rutaceae, Lauraceae, and Ericaceae.

Of the genera identified, 56 have cosmopolitan distributions, while 652 are non-cosmopolitan, mainly tropical and subtropical. Unique ecological phenomena such as stem flowering, strangulation, and aerial epiphytism are frequently observed in species like *Ficus*, *Artocarpus*, and *Drynaria roosii*. Tropical elements constitute 68.6% of the genera, while temperate elements make up 31.4%, highlighting the transitional nature of the flora. Many genera, including *Endospermum chinense* and *Castanopsis jucunda*, reflect the northward diffusion of subtropical flora along the Pearl and North River basins (Liao 1992; Li 2018).

Geographically, Mount Danxiashan's flora is a subset of the Nanling Mountains flora, shaped by its distinctive red bed landform and characterized by xerophytic species on cliffs, wetland populations, and valley monsoon forests. Noteworthy species include Alsophila spinulosa, *Ophioglossum vulgatum*, *Angiopteris fokiensis*, and tropical woody vines like *Gnetum* and *Mucuna*. The Danxia flora uniquely integrates ecological and geographical features, earning it the designation of Geo-eco-geographical flora.

These findings underline Mount Danxiashan's ecological importance and contribute to a deeper understanding of its biodiversity and conservation value.

3.2 Endemism of Flora in Mount Danxiashan

The defining feature of the danxia landscape is its extensive and widely distributed cliffs. These cliffs have given rise to isolated, island-like peaks and enclosed valleys, creating highly diverse ecological niches that support the preservation of numerous regional endemic species. Composed primarily of red glutenite, siltstone, and argillaceous rock, the cliffs are characterized by poor water retention. Rapid erosion and vegetation regeneration act as natural filters, enabling only certain plants to thrive. Furthermore, the hundreds of isolated peaks at Mount Danxiashan create

Fig. 2 Some new species found in Mount Danxiashan. **a** *Diospyros danxiaensis*; **b** *Primulina danxiaensis*; **c** *Primulina huangjiniana*; **d** *Danxiaorchis singchiana*; **e** Viola *hybantoides*; **f** *Semiaquilegia danxiashanensis*; **g** and **h** *Lespedeza danxiaensis*; **i** *Spiradiclis danxiashanensis*; **j** *Selaginella orientali-chinensi*; **k** and **m** *Phyllostachys danxiashanensis*; **n** *Dendrocalamus pulverulentoides*; **o** *Dryopteris shiakeana*; **p** *Asplenium danxiaense*

significant barriers to gene flow between species, fostering the differentiation of new species. This close interplay between endemism, geology, and lithology is a hallmark of Mount Danxiashan's flora.

The region is home to one family endemic to East Asia, Actinidiaceae, and 12 genera endemic to China, including *Cunninghamia*, *Primulina*, *Semiliquidambar*, *Danxiaochis*, *Thyrocarpus*, *Pteroceltis*, *Eurycorymbus*, *Petrocodon*, *Schnabelia*, and *Tetrapanax*. These genera underscore the antiquity and botanical richness of the Nanling Mountains flora.

Mount Danxiashan also harbors a variety of locally endemic species, such as *Firmiana danxiaensis*, *Diospyros danxiaensis*, *Primulina danxiaensis*, *Danxiaorchis singchiana*, *Viola hybantoides*, *Semiaquilegia danxiashanensis*, *Lespedeza danxiaensis*, *Selaginella orientali-chinensi*, *Phyllostachys danxiasanensis*, and *Dendrocalamus pulverulentoides*. Many of these species are uniquely adapted to the siltstone, sandstone, and conglomerate cliffs of the area, displaying narrow habitat endemism. These features highlight the ecological and evolutionary significance of Mount Danxiashan's unique flora (Shen et al. 2010; Wang et al. 2015) (Fig. 2).

3.3 Rare and Endangered Plants in Mount Danxiashan

According to the List of National Key Protected Wild Plants (2021), there are about 30 species of national key protected wild plants in Mount Danxiashan, including 1 species under level I, namely *Cycas fairylakea*, and 29 species under level II. According to the statistics of the China's Red List of Biodiversity-Higher Plants Volume (2017), there are a total of 44 species of rare and endangered plants (Table 1).

4 Fauna

Fieldwork and literature reveal that Mount Danxiashan is home to 1,426 animal species, spanning 255 families across 47 orders.

Table 1 List of rare and endangered wild plants in Mount Danxia (Wu 1980, 2003)

No.	Families	Species	List of national key protected wild plants	China's red list of biodiversity
1	Cycadaceae	*Cycas fairylakea*	I	CR
2	Lycopodiaceae	*Huperzia serrata*	II	EN
3	Marattiaceae	*Angiopteris fokiensis*	II	
4	Dicksoniaceae	*Cibotium barometz*	II	
5	Cyatheaceae	*Alsophila spinulosa*	II	
6	Cyatheaceae	*Gymnosphaera podophylla*	II	
7	Pteridaceae	*Ceratopteris thalictroides*	II	VU
8	Aristolochiaceae	*Asarum insigne*	II	VU
9	Aristolochiaceae	*Aristolochia fordiana*		VU
10	Podocarpaceae	*Podocarpus macrophyllus*	II	VU
11	Orchidaceae	*Anoectochilus roxburghii*	II	EN
12	Orchidaceae	*Bletilla striata*	II	EN
13	Orchidaceae	*Cymbidium ensifolium*	II	VU
14	Orchidaceae	*Cymbidium goeringii*	II	VU
15	Orchidaceae	*Cymbidium kanran*	II	VU
16	Orchidaceae	*Danxiaorchis singchiana*	II	
17	Orchidaceae	*Dendrobium loddigesii*	II	VU
18	Orchidaceae	*Dendrobium officinale*	II	
19	Orchidaceae	*Tainia ruybarrettoi*		EN
20	Orchidaceae	*Goodyera yangmeishanensis*		VU
21	Orchidaceae	*Nervilia plicata*		VU
22	Poaceae	*Sorghum propinquum*	II	EN
23	Fabaceae	*Glycine soja*	II	
24	Fabaceae	*Spatholobus suberectus*		VU
25	Fabaceae	*Ormosia henryi*	II	VU
26	Fabaceae	*Ormosia semicastrata*	II	
27	Fabaceae	*Ormosia semicastrata* f. *pallida*	II	
28	Theaceae	*Camellia sinensis* var. *pubilimba*	II	
29	Rutaceae	*Fortunella japonica*		EN
30	Rutaceae	*Citrus mangshanensis*	II	
31	Rutaceae	*Fortunella hindsii*	II	
32	Rutaceae	*Fortunella venosa*	II	
33	Schisandraceae	*Kadsura coccinea*		VU
34	Pentaphylacaceae	*Ternstroemia conicocarpa*		VU
35	Sapindaceae	*Dimocarpus longan*	II	VU
36	Sapindaceae	*Eurycorymbus cavaleriei*	II	
37	Hydrocharitaceae	*Ottelia alismoides*		VU
38	Dioscoreaceae	*Dioscorea persimilis*		EN
39	Moraceae	*Artocarpus hypargyreus*		EN
40	Rosaceae	*Rosa kwangtungensis*	II	VU
41	Ixonanthaceae	*Ixonanthes reticulata*		VU
42	Malvaceae	*Firmiana danxiaensis*	II	CR
43	Pittosporaceae	*Pittosporum fulvipilosum*		VU
44	Pteridaceae	*Adiantum juxtapositum*		VU

4.1 Insects

A total of 1,023 species from 783 genera in 176 families and 16 orders have been identified. *Lepidoptera* and *Coleoptera* dominate, comprising more than half of the recorded insect species, while aquatic beetles are relatively scarce. About 20% of the insects are endemic to China, including species unique to the Nanling Mountains or Guangdong, such as *Toacris yaoshanensis*, *Flatocerus nankunshanensis*, *Carabus* shun floridus, and *Carabus melli*. Three species, including *Cheirotonus jansoni*, are listed in the China Species Red List, with *Cheirotonus jansoni* designated as a Level II national key protected species.

4.2 Fish

Forty species or subspecies, belonging to 12 families and 4 orders, have been identified, with 90% holding significant economic value. Endemics include *Microphysogobio kachekensis* and *Schistura rara*. Six species, such as *Parasacco spilurus* and *Oryzias latipes*, are listed in the China Species Red List.

4.3 Amphibians

There are 29 species from 20 genera in 7 families, with one, *Hoplobatrachus rugulosus*, listed as a Level II national key protected species. Endemics to China include *Megophrys mangshanensis* and *Rana longicrus*. *Six amphibians* appear on the China Species Red List.

4.4 Reptiles

Seventy-two species from 48 genera in 21 families have been recorded, with eight, such as *Cuora trifasciata*, listed as Level II national key protected species. Endemics include *Sacalia bealei* and *Eumeces chinensis*, while 18 species are listed in the China Species Red List. The endangered *Platyplacopus sylvaticus* has also been discovered here.

4.5 Birds

The area supports 236 bird species from 60 families in 18 orders, including two Level I and 37 Level II national key protected species, such as Lophura nychthemera. Endemic species include *Bambusicola thoracica* and *Spizixos semitorques*. Five species, including *Passer montanus*, appear on the China Species Red List.

4.6 Mammals

Forty-seven mammal species from 16 families in 6 orders inhabit the region. The fauna shows a blend of subtropical and tropical characteristics, with some northern species like *Mustela sibirica*. Fourteen species are listed as national key protected wild animals, including 3 at Level I: *Neofelis nebulosa*, *Moschus berezovskii*, and *Muntiacus crinifrons*. Eight mammals are endemic to China, such as *Mogera insularis* and *Rhinolophus rouxi*, while 54 species are classified as endangered in China, including 1 critically endangered, 8 endangered, and 45 vulnerable.

Mount Danxiashan demonstrates a rich and diverse array of species, highlighting its ecological importance and the need for ongoing conservation efforts.

5 Nature Conservation

5.1 Development History of Mount Danxiashan Nature Reserve System

In 1995, Mount Danxiashan received approval from the State Council to establish a national nature reserve. This marked the beginning of a series of protection and management initiatives, including the introduction of relevant laws and regulations to ensure the effective preservation of the area's geological, geomorphological, and ecological environments.

In 2010, under the joint declaration of 'China Danxia', Mount Danxiashan, alongside Mount Langshan (Hunan), Mount Taining (Fujian), Mount Chishui (Guizhou), Mount Longhu (Jiangxi), and Mount Jianglang (Zhejiang), was designated a World Natural Heritage site by the International Union for Conservation of Nature.

Building on this foundation, national and local governments have steadily enhanced conservation efforts and ecological development in Mount Danxiashan. Central to these efforts is the 'Overall Plan for Danxiashan National Park (2017–2035)', which emphasizes balancing protection with development. The plan focuses on preserving the area's natural ecological environment while maintaining the authenticity and integrity of its natural and cultural heritage.

The strategy also highlights the need to standardize park management, elevate the scientific and cultural value of resource utilization, improve environmental quality and infrastructure, and ensure sustainable resource use. A key priority is striking a balance between the livelihoods of the park's surrounding residents and its ecological preservation to promote regional economic growth and social progress.

Currently, Mount Danxiashan is listed in the preliminary plan for national park construction by the National

Forestry and Grassland Administration. In the coming years, the natural protection and ecological development of Mount Danxiashan are expected to gain momentum, further solidifying its status as a model for ecological civilization and sustainable growth.

5.2 Protection and Enhancement of Ecological Environment in Danxia Landscape

5.2.1 Protection of Danxia Landscape and Vegetation

Priority protection areas should be designated based on geomorphic types, habitat types, vegetation types, and ecosystem types. Specific protection measures must be developed for communities hosting rare and endangered species. Similarly, efforts should focus on small geomorphic and habitat populations and endemic species such as the *Cycas fairylakea* population, Wutong Danxia population, Mangshan wild citrus population, white osmanthus population, Dabao white mountain tea population, Suihua fir population, and *Alsophila spinulosa* population.

5.2.2 Protection of Wetlands and River Basins

The North River basin is crucial for the exchange and diffusion of Mount Danxiashan's flora and fauna, serving as an essential ecological barrier in northern Guangdong. Its water quality directly impacts the safety of urban drinking water. Within the Danxiashan Nature Reserve, the Jinjiang and Zhenjiang Rivers and other reservoirs are managed under the river and lake chiefs' systems, implementing comprehensive river governance. Upstream sewage outlets have been standardized, while sewage treatment facilities and pipelines have been upgraded.

Recent water quality monitoring in Renhua County confirms that all indicators meet Class III standards, with a 100% compliance rate for centralized drinking water sources and surface water sections. Improved water quality has significantly enhanced biodiversity, with notable increases in fish and bird species.

5.2.3 Protection of Biodiversity and Promotion of Ecological Sustainability in Communities

1. Protection of ancient trees

Ancient trees hold critical economic, ecological, landscape, and research value. Approximately 800 ancient trees, representing 36 species across 18 genera and 20 families (e.g. Lauraceae, Fagaceae, Camelliaceae), are distributed in Mount Danxiashan. Of these, 34 are classified as Class I (over 300 years old) and over 760 as Class II (100–290 years old). Current protective measures include:

- Strict penalties for logging, transplanting, damaging, or trading ancient trees.
- Compiling an ancient tree registry and assigning responsibility to individuals or organizations.
- Raising public awareness through media campaigns and highlighting the historical, cultural, and landscape value of ancient trees.
- Conducting regular inspections to address issues like decay, pests, and dead branches.

2. Protection of rare and endangered species

Rare and endangered plants are invaluable for understanding species origins, evolution, genetic breeding, and historical geology and climate. A survey identified 44 key rare and endangered species in Mount Danxiashan. Measures for their protection include defining specific protected areas, categorizing landform types, enforcing wildlife regulations, creating wildlife corridors, conducting artificial breeding research, and promoting forest and wildlife conservation laws.

3. Community involvement and sustainable use of ecological resources

Mount Danxiashan is rich in plant germplasm resources, serving as a critical resource bank. Ecological protection initiatives include:

- Protecting wild plant resources and managing scenic areas under strict regulations to minimize tourism's environmental impact. Implementing robust monitoring to prevent illegal logging and forest fires.
- Developing medicinal plants while balancing pharmaceutical industry growth with environmental sustainability.
- Conducting scientific research on resource plants, cultivating rare and economically valuable species, and assessing ecotourism's impact on vegetation.
- Encouraging community participation and awareness campaigns to promote sustainable resource use, including compiling a medicinal plant species list.

By adopting these measures, Mount Danxiashan ensures the long-term sustainability of its ecological and cultural heritage while promoting harmonious coexistence between conservation and community development.

6 Conclusions

The development of the tourism industry, particularly the construction of facilities and infrastructure, has resulted in pollution and visual impacts on the local water environment. Additionally, it has disrupted animal habitats and undermined the integrity and stability of the ecosystem, posing significant threats to biodiversity conservation. Concurrently, the outdated production methods and lifestyles of some impoverished farmers, combined with the overexploitation of natural resources in surrounding areas driven by rapid socio-economic growth, have exerted considerable pressure on environmental protection. Striking a balance among ecological conservation, agricultural and forestry production, and tourism development remains a pressing and complex challenge.

References

Fan Q, Chen S, Wang L et al (2015) A new species and new section of *Viola* (Violaceae) from Guangdong, China. Phytotaxa 197(1):15–26

Li DZ (2018) Dictionary of families and genera of vascular plants in China. Science Press, Beijing

Liao WB (1992) Subtropical flora of Guangdong. Ph.D. dissertation, Sun Yat-Sen University

National Forestry and Grassland Administration & Ministry of Rural Affairs and Agriculture (2021) List of national key protected wild plants. Beijing

Niu ZY, Tong YH, Ni JB et al (2020) *Dendrocalamus pulverulentoides* (Poaceae: Bambusoideae), a new bamboo species from Guangdong, China. J Bamboo Res 39(3):34–39

Peng SL, Liao WB, Li Z et al (2011) Comprehensive scientific investigation of animal and plant resources in Mount Danxiashan, Guangdong. Science Press, Beijing

Qin HN, Zhao LN et al (2017) Red list of China higher plants. Biodiversity 25(7):689–695

Shen RJ, Lin SS, Yu Y, Cui DF, Liao WB (2010) Chiritopsis danxiaensis sp. nov. (Gesneriaceae) from Mount Danxiashan, south China. Nordic J Bot 28(6):728–732

Wang R, Wen H, Deng S et al (2015) *Spiradiclis danxiashanensis* (Rubiaceae), a new species from South China. Phytotaxa 206(1):30–36

Wu ZY (1980) Vegetation of China. Science Press, Beijing

Wu ZY, Zhou ZK, Li DZ et al (2003) The areal types of seed plant families in the world. Acta Bot Yunnanica 25(3):245–257

Zheng X, Tong Y, Ni J, Chen Z, Xia N (2019) *Phyllostachys danxiashanensis* (Poaceae: Bambusoideae), a new species from south China. Phytotaxa 388(2):201–206

Zhou JJ, Huang ZP, Li JH, Hodges S, Deng WS, Wu L, Zhang Q (2019) *Semiaquilegia danxiashanensis* (Ranunculaceae), a new species from Danxiashan in Guangdong, southern China. Phytotaxa 405(1):1–14

History and Culture

Rongfeng Hou

Abstract

Mount Danxiashan boasts a profound cultural legacy, woven into its history. This iconic mountain has long been a cradle of cultural and religious significance, captivating the hearts and minds of many. Its rich heritage is reflected in the numerous ancient temples and historical sites scattered across the landscape, each telling a unique story of devotion, perseverance, and artistry.

1 Overview

Mount Danxiashan boasts a profound cultural legacy, woven into its history. This iconic mountain has long been a cradle of cultural and religious significance, captivating the hearts and minds of many. Its rich heritage is reflected in the numerous ancient temples and historical sites scattered across the landscape, each telling a unique story of devotion, perseverance, and artistry (Fig. 1).

1.1 History

During the Sui and Tang dynasties (581–907 AD), Mount Danxiashan was already a famous scenic spot in the Lingnan (South China) region. Monks and nuns entered the mountains to practice and built Buddhist temples. By the Chongning era of Emperor Huizong in the Northern Song Dynasty (1102–1106), the lay Buddhist Fayun traveled to Mengjue Pass in Danxiashan. Upon seeing the extraordinary caves and breathtaking landscapes, where the rocks' colors resembled rich cinnabar and shone like radiant clouds, he experienced a sudden spiritual awakening. He inscribed the words 'Mengjue Pass' (Dream Awakening Pass) and built an 18-room hall there, enshrining Guanyin Bodhisattva (Goddess of Mercy).

1.2 Seclusion and Donation

In the final years of the Ming Chongzhen era (1644), Li Yongmao, the Governor of Ganzhou Jiangxi, retreated to Mount Danxiashan's 'Zhanglao Village' with his family after failing to resist the Qing forces. He built homes, cultivated fields, and hosted gatherings with friends. Later, his brother Li Chongmao donated the mountain to a Buddhist temple.

1.3 Flourishing Buddhism

In the first year of the Kangxi era in the Qing Dynasty (1662), Chan Master Dangui from Guangzhou's Haichuang Temple arrived at Danxia Mountain to establish a Buddhist sanctuary, founding Biechuan Temple. Over time, he constructed the Great Buddha Hall, Maitreya Hall, Guanyin Pavilion, Scripture Repository, Abbot's Quarters, meditation rooms, and guest halls, creating a sizable mid-mountain monastery. The name 'Biechuan Temple' derives from the Chan Buddhist teaching: 'Not relying on written words, a special transmission outside the scriptures, directly pointing to the human mind, realizing one's nature to attain Buddhahood'.

1.4 Changes Over Time

From the establishment of sacred monasteries to the influence of renowned scholars and monks, Mount Danxiashan's cultural heritage is a testament to the enduring spirit of the

R. Hou (✉)
Danxiashan Adminstrative Committee, Danxiashan UNESCO Global Geopark, Shaoguan City, Guangdong Province, China
e-mail: hspdz@163.com

Y. Ng and W. Eder (eds.), *The Redstone Park of China*, Geoheritage, Geoparks and Geotourism,
https://doi.org/10.1007/978-3-031-95603-4_6

Fig. 1 Sunrise in Mount Danxiashan. It is one of the best spots to view sunrise in South China

region. The mountain's picturesque scenery and its historical and spiritual significance continue to inspire visitors and pilgrims alike.

Mount Danxiashan, along with Mount Luofu, Mount Dinghu, and Mount Xiqiao, is known as one of the 'Four Famous Mountains South of the Five Ridges' in China. Renowned for its stunning danxia landscape, Mount Danxiashan has a rich history for reference.

1.4.1 Early Historical Context

Prior to the Sui Dynasty (581 AD), Mount Danxiashan remained largely undocumented, functioning as a natural barrier with minimal cultural imprint. Archaeological insights suggest that the broader Shaoguan region witnessed late Neolithic activity (ca. 4,000–50,000 years BP), as evidenced by the Shixia Culture site. Although Danxiashan's caves may have served as temporary shelters, direct habitation evidence is lacking. Following Qin unification (214 BC), the area fell under successive administrative entities—first Nanhai, then Guiyang Commandery—yet its rugged terrain likely deterred major settlement. During the Three Kingdoms and Northern and Southern Dynasties (220–581 AD), it was gradually absorbed into Shixing Commandery. Despite the spread of Buddhism into Lingnan, there are no verified temple records predating the Tang Dynasty; legends of Jin Dynasty monastics remain anecdotal. References to Bodhidharma's journey through nearby Shaozhou (now called Shaoguan) exist, though his association with Danxiashan is unconfirmed. The region's early significance was thus defined more by topography than by cultural or political centrality. During the Qing Era (1662–1911), Jinshiyan remained active throughout the period and Biechuan Temple was established in 1662.

The early development of Mount Danxiashan is further discussed particularly in Sect. 1.1 in Chap. 7.

1.4.2 Modern Historical Context

Between 1912 and 1949, the area avoided major wartime damage due to remoteness. During the period 1920s–40s, Geologist Feng Jinglan of Sun Yat-sen University studied danxia landforms and laid groundwork for later research. In the early years of the People's Republic of China (1949–1963), Biechuan Temple repurposed for secular use; surrounding forests briefly logged during the Great Leap Forward. Its modern history after 1962 can be summarized as follows:

- 1963: Establishment of Mount Danxiashan Forest Farm.
- 3 January 1980: Opening to foreign tourism.
- 1984: Xi Zhongxun (father of President Xi Jinping) inscribed the name for the old gate of 'Mount Danxiashan'.

- 1988: Approved by the State Council and listed as a national scenic spot.
- 1995: Designated as a national nature reserve.
- 2004: Recognized by UNESCO as one of the first global geoparks parks.
- 2010: Included in the World Heritage List under 'China Danxia', representing seven concentrated distribution areas of Danxia landscape in six provinces of southern China.
- 2011: Awarded the status of a national 5A tourist attraction in China.
- Currently, efforts are underway to incorporate Mount Danxiashan into the National Park System of China.

Mount Danxiashan is situated at the southern foot of the Nanling (Southern) Mountains and the upper reaches of the Beijiang (North) River System in the Pearl River Basin. The mountain's rocks are named for their striking red color and cloud-like brightness. As a prominent representative of low-altitude mountain-type scenic spots in China, Mount Danxiashan features a significant concentration of danxia landforms from the Mesozoic era.

According to geomorphologists, among the more than 1,100 danxia landscapes discovered in 29 provinces (including municipalities, autonomous regions, and special administrative regions) across China, Mount Danxiashan stands out as the most developed, most complete in type, richest in shape, and most scenic site (Huang 2009; Chen and Liu 2008; Guo et al. 2022; Fang et al. 2022). It is the most beautiful and well-studied area for landforms with danixa characteristics. Comparable to Yellowstone National Park in the United States, it is a mountain and park affectionately known as China's 'Red Stone Park'.

The modern and latest development of Mount Danxiashan is further discussed particularly in Sect. 1.2 in Chap. 7.

1.5 Iconic Landscape of Guangdong Province

Mount Danxiashan spans a planned total area of 292 km^2 and features over 680 red sandy conglomerate rocks with a relative height exceeding 50 m. Characterized by its striking red cliffs, the area is notable for its geological beauty and cultural heritage.

The mountain is home to more than a dozen temple ruins, including Biechuan Temple, Jinshiyan Temple, Xianjuyan Taoist Temple, Jinguiyan Temple, and Jinlongshanyan Temple, as well as over 100 ancient fortresses. Throughout history, literati and scholars from various dynasties have left thousands of poems and essays here. Additionally, there are more than 300 inscriptions on the cliffs.

The immovable cultural relics of Mount Danxiashan encompass ancient temples, cottages, villages, post roads, cliff carvings, petroglyphs, rock coffins, and tombs. Notably, the Mount Danxiashan Cliff Stone Carvings are recognized as the national key cultural relic protection units. Other significant sites include the Neolithic Catfish Zhuan Site, Ximei Village from the Ming and Qing dynasties, Xueyan Temple and Shiruquan from the Qing dynasty, the Lion Rock Temple Site from the Ming and Qing dynasties, and the Mount Danxiashan Pagoda Tomb Group from the Qing dynasty. These five sites are designated as cultural relic protection units in Guangdong Province.

The foundation of Mount Danxiashan is an intermountain lacustrine sedimentary basin dating back to the Cretaceous period, approximately 140–65 million years ago. Since around 65 million years ago, the lacustrine red sandy conglomerate layer, which is 4,000 m thick and covers an area of 500 km^2, has been uplifted almost horizontally due to the Yanshan and the Himalayan Movements.

During this uplift process, the rock formations experienced varying degrees of fracturing, leading to the development of endogenous interlayer and cross-layer joints. The uplifted rock formations were heavily eroded by flowing water, which, along with differential weathering and gravity collapse over the years, has sculpted the landscape.

Since the Quaternary period, the distinctive features of Mount Danxiashan have become increasingly prominent, resulting in a natural sculpture park. The exquisite beauty of the danxia landscape can be likened to a shrunken fairyland and an enlarged bonsai.

2 Origin of Names: Danxia and Danxiashan

2.1 Exploring the Characters Dan and Xia

Dan (丹): This is a pictographic Chinese character. 'Cong yi' refers to the horizon; 'Cong jiōng' also indicates the horizon, symbolizing the sky that envelops everything. 'Cong ri' represents the sun. Together, these elements depict a red sun slowly rising from the horizon or gently setting into the sky.

Xia (霞): This character signifies the beautiful illusion seen after the rain under the sunlight. The morning glow and the sunset glow are particularly enchanting, capturing the essence of radiant beauty (Huang 2009; Ai and Zhang 1994; Hou 2019).

2.2 The Meaning of Danixa

Danxia is a literary term that refers to the red, colourful clouds seen during sunset or sunrise. The earliest known reference can be found in Cao Pi's 'Furong Pond Zuo' from the Three Kingdoms period (220–280 AD). 'Danxia embraces the moon, and Huaxing rises from the cloud'. This line poetically describes a bright moon nestled among beautiful clouds, with Venus (known as Qiming in the morning and Changgeng in the evening) twinkling within—a vivid image of colourful clouds, stars, and the moon.

The term 'Danxia' was first used to describe Mount Danxiashan in Lun Yiliang's 'Jin Shiyan' during the Ming Dynasty: 'When the water is exhausted and the rocks are seen on the cliffs, Danxia is between the Han and Han. Nuwa's five-coloured stones, Yushun's thousand-year-old mountain'. In this context, 'Danxia' remains a literary term.

2.3 Shaoshi

Mount Danxiashan was historically known as Shaoshi. As Qu Dajun from the Qing Dynasty noted: 'Mostly, the mountains in Shao are strange, but Shaoshi is the most. If there are two towers, Shaoshi is the most' (Ai and Zhang 1994). According to legend, Emperor Shun visited Shaoshi in the south four thousand years ago to play Shao music and bring order to Shaoshi. The names Shaozhou and Shaoguan are derived from Shaoshi, which in turn was named after Shaole.

Until the Great Ming Revolution, brothers Li Yongmao and Li Chongmao purchased Zhanglaozhai by the Jinjiang River to seek refuge and named it Mount Danxiashan. Shaoshi has always been the cultural heart of the Mount Danxiashan. Li Daoyuan of the Northern Wei Dynasty wrote in the 'Shui Jing Zhu Qin Shui: 'Two stones face each other, one mile apart, roughly equal in size, resembling double towers, and they are called Shaoshi'. The palace gate is called Shaoshi. This indicates that Shaoshi became a scenic spot in South China as early as the Wei and Jin Dynasties.

The 'Tang Yuanhe County Chronicles' states: 'In the ninth year of Sui Kaihuang (589 AD), Chen was flattened, Donghengzhou was renamed Shaozhou, and the name was taken from Shaoshi to the north of Zhou'. Emperor Wen of the Sui Dynasty pacified the Chen Dynasty in 589 AD, unifying China, and renamed Donghengzhou to Shaozhou after Shaoshi.

Shaoshi is primarily located on the right bank of the lower reaches of the Zhenjiang River in the Beijiang water system within the Pearl River Basin, known as the 'Thirty-six Stones of Shaoshi'. These include 36 famous stones, such as the double quers resembling palace gates, the typical wall-shaped landforms of Danxia, and other notable features like the gate stone of a Chinese football goal from the Song Dynasty with a ball boy standing on the west side, the rock pillar daughter-in-law stone for women, the lion stone for lions, and the Wuyang stone for five sheep (Fig. 2).

Fig. 2 The history of Mount Danxiashan is etched into its rock surfaces

Fig. 3 The picturesque landscape of Mount Danxiashan

2.4 The Naming of Mount Danxiashan

Traditionally, Mount Danxiashan referred only to the Zhanglaofeng area. The mountain was named by brothers Li Yongmao and Li Chongmao from Dengzhou, who purchased the mountain and chose to live in seclusion there in the second year of Longwu in Nanming (autumn in 1646 AD). In Li Chongmao's 'Mount Danxiashan Records', it is written: 'One night, Zhou Zichanggong, a student of Yitai, and Liu Ziwangfu, Songtao Kunyu, a literary student, had a night talk with Mingwan. They talked about twenty miles south of Zhinan, and there was Mount Danxiashan' (Hou 2019).

In the early Qing Dynasty, Chen Shiying wrote in the 'Danxia Shanshui General Preface': 'It may be said that Danxia is the old place of burning wood Buddha, and it is not appropriate to change its name. It is also a native of Dengzhou, Nanyang, Li Gong who took Mount Danxiashan to show that he can't bear to forget his origin.

The story of 'burning wood Buddha's old place' refers to the Zen master Danxia from Dengzhou during the Tang Dynasty (618–907 AD), who once used a wooden Buddha statue to warm himself in winter. In 1646 (Bingxu year), during the Zhongyang Festival in autumn, brothers Li Yongmao and Li Chongmao bought the mountains and forests around Zhanglaofeng from the Liu family in Renhua County to escape chaos and live in seclusion. They borrowed the name of Mount Danxiashan from Nanyang, Henan, and named the rocks around Zhanglaofeng in Guangdong as Mount Danxiashan (Fig. 3).

3 Seven Scenic Spots and Ten Cultural and Scientific Study Routes

Mount Danxiashan is composed of seven major areas: Danxia Scenic Area, Bazhai Scenic Area, Shaoshi Scenic Area, Feihuashui Scenic Area, Xiamenji Scenic Area, Jinjiang Scenic Belt, and Zhenjiang Scenic Belt. It is renowned for its natural wonders, the beauty of Jinjiang, the seclusion of Jinyan, the charm of Shaoshi, and the richness of Xiafu. Currently, the accessible areas are divided into ten scientific research study routes, allowing visitors to explore and study the not only the unique geological but also the cultural features of the area (Danxiashan, 2024) (Fig. 4).

3.1 Route 1—Jinshiyan

The Zhanglaofeng Tourist Area is composed of upper, middle, and lower landscapes. The middle and lower layers are designated as part of the Mount Danxiashan Scientific Research Jinshiyan Route 1. Jinshiyan connects to

Fig. 4 The temple of Jinshiyan was built inside a cave over a thousand years ago

Hailuoyan (Conch Rock) at the top and the Jinjiang River below. It features four large horizontal caves: Qianshengyan, Zushiyan, Fuhuyan, and Longwangyan. Other notable formations include Longlinpianshi (Dragon Scale Flake Stone), Baizhangxia Yixiantian Gorge,

Jinyanchibi (Jinyan Red Cliff), Mawei (Horse tail) Waterfall, and the 'Jinyan' Cliff Group.

Jinyan Cliff, a prominent feature of the danxia landscape, boasts a vertical height of 200 m. From bottom to top, it showcases a complete geological section of the Bazhai Member and Jinshiyan Member of the Mount Danxiashan Formation. The Yuquanchi Pool, with its three sides, serves as a drinking water source for monks and visitors in Jinyan. The tour duration for this line is approximately 2–3 h (Fig. 5).

3.2 Route 2—Zhanglaofeng (Elder's) Peak

From the entrance of Zhanglaofeng via Banshan Pavilion to the middle and upper levels, this route is designated as Route 2 of Zhanglaofeng in Mount Danxiashan. Zhanglaofeng (Elder's Peak) stands at 351 m above sea level, with Baozhu Peak, the highest point, reaching 409 m above sea level. Visitors can access these peaks via the Danti Iron Lock, Danxia Plank Road, or by cable car.

In the middle layer, you will find scenic spots such as the 'Danxia' Cliff Group, Biechuan Temple, Gospel Gorge, Mandarin Duck Tree, Natural Rock, Yufeng Pavilion, and Shuangxi Terrace. The upper layer features Zhanglaofeng Peak, Conch Peak, and Baozhu Peak standing side by side, surrounded by large horizontal caves such as Xueyan, Wanxiuyan, Shuilianyan, Longwangyan, and Schistyan.

Viewing spots like Zhanglaofeng Riting Pavilion, Shaoyin Terrace, and Rudder Stone provide the best vantage points to fully appreciate the danxia landscape and peaks, as well as to witness the sunrise and sunset. From the top of the mountain, you can overlook Mitra Peak, Wangfu Stone, Tianzhu Stone, Jinjiang, Bazhai, Teapot Peak, Sister Peak, and Sleeping Beauty. The tour duration for this route is approximately 2–4 h (Fig. 6).

3.3 Route 3—Xianglonghu (Flying Dragon) Lake

From the entrance of Zhanglaofeng Peak, around Xianglonghu Lake, through Yinyuan Stone and Mushroom Pavilion, and back to Xianglonghu Lake, this route is designated as the Danxia Mountain Scientific Research Xianglong Lake-Yinyuan Stone Route 3. Xianglonghu Lake, constructed in 1984, features a depth of 20 m, a length of over 1,300 m, a surface area of 10 hectares, and a storage capacity of 580,000 m^3. The lake is named for its outline, which resembles a soaring green dragon.

Along the lake, visitors can explore scenic spots such as Longxujian, Pojunzhai, Xianjuyan Taoist Temple, and Panda Stone. Approximately 700 m from the end of the lake lies Yinyuanshi (Female Rock), a natural stone cave resembling female reproductive organs. The stone measures 10.3 m in height, 4.8 m in width, and 4.3 m in depth. The tour duration for this route is approximately 2–3 h.

Fig. 5 Viewing platform in Zhanglaofeng, the Elder's Peak

Fig. 6 The fallen boulder from the cliff in Xianglonghu Lake provided a rock craving surface to mark and record the local culture and belief

3.4 Route 4—Wolonggang (Lying Dragon) Hillock

From the end of Xianglong Lake, through Butterfly Valley, Wolonggang, Huangshakeng, Baota Peak, Yuping Peak, and back to the end of Xianglonghu Lake, this route is designated as Mount Danxiashan Scientific Exploration Wolonggang Route 4. Wolonggang is a small hill situated in the heart of the Danxia peak forest. From here, you can admire the surrounding peaks of Elder, Pagoda, Monk Hat, Yangzhou, and Pojun.

The route delves deep into the hinterland of Mount Danxiashan, featuring dramatic danxia walls, intersecting valleys, and an exceptional ecological environment. It offers the perfect opportunity to breathe in the fresh air and fully appreciate the danxia landscape and the region's biodiversity. The tour duration for this route is approximately 4–6 h.

3.5 Route 5—Yangyuanshi (Male Rock)

This inspection route highlights the Danxiayan pinnacle, Yangyuanshi Rock (Male Rock), as the central landscape feature. Ximei Village, sitting at an elevation of 286 m, is a typical danxia rock wall formation. At the east end of Ximeizhai, Yangyuan Stone stands tall and straight, resembling a male reproductive organ. This natural rock pillar, 28 m high, 7 m in diameter, and weighing 600 tons, is famously known as 'world's weirdest rock'.

The Worshiping Balcony offers the best vantage point for viewing and admiring Yangyuanshi Rock. From here, you can take the steep Cloud Cliff Plank Road to the mountain's summit. At Jiadun Pavilion, the panoramic view reveals the winding Jinjiang River and Zhanglaofeng (Elder's Peak), the main mountain of Mount Danxiashan, in full clarity. The summit features Ximei Village, established in the late Ming Dynasty (1368–1644 AD). One can access it via the Yunya Plank Road and the Jiujiu Tian Ladder. The tour typically takes 2–3 h.

3.6 Route 6—Tongtai Bridge

This route relates to the Yangyuanshi Line 5 and is highlighted by the natural arch named Tongtai Bridge. Aptly described as 'One bridge flying across the north and south', the Tongtai Bridge serves as the central landscape feature. Attractions along this route include Tongtai Bridge, Sunbuyan (Cloth Hanging Rock), Haibaoshi (Seal Rock), the ruins of Shiziyan (Lion Rock) Temple, and more.

The Tongtai Bridge, stretching from east to west, boasts a bridge deck 48 m long, 4.5–8.0 m wide, and 3–5 m thick. With an inner span of 35 m and a clearance of 13 m, it is renowned as 'China's No. 1 Danxia Bridge' and is considered a typical, natural arch of danxia landform. The Haibaoshi (Seal Rock) resembles a seal, complete with a leopard head, body, and whiskers. The tour typically takes 2–3 h.

3.7 Route 7—Water Danxia

Jinjiang River, also known as Renhua River, originates from Immortal Ridge in Zhudong, Chongyi County, Jiangxi Province, and spans a total length of 109 km. The 34 km downstream section of the river forms an embedded endogenous meander that flows through Mount Danxiashan from north to south. The river is divided into upper, middle, and lower sections by the Danxiashan Power Station and the Yaoshan Power Station.

Jinjiang River connects scenic spots like Danxiashan, Bazhai, and Shaoshi in a seamless manner. Notable among the landscape features are six deep meanders, including Changshabei, Niubi, and Nanyangeng, winding elegantly through the mountains. The section of the river between Huangwu Hydropower Station and Danxia Hydropower Station is designated as the Water Danxia Line 7 Cruise Ship Route, with a total length of 7 km.

Along this route, you can enjoy picturesque sights such as Sleeping Beauty, Elephants Coming Out of the Mountain, Carp Leaping Over the Dragon Gate, Golden Turtle Pilgrimage, Six Fingers Capturing Demons, Ancient Quarry, Guanyinshi Rock, and Teapot Peak. The tour takes approximately 1 h.

3.8 Route 8—Jinjiang Gallery

This line is situated between Danxiashan Power Station and Yaoshan Power Station, encompassing the area from Changshabei Deep Meander to Niubi Deep Meander, with a total length of 7 km. Along the route, visitors can admire scenic spots such as Sister Peak, Thumb Rock, Bazhai, Teapot Peak, Shangtianlong, Guanyinshi Rock, and Tianluo Stone. The tour typically takes 1 h (Fig. 7).

3.9 Route 9—Bazhai

The Bazhai Scenic Area is in the western part of Mount Danxiashan, controlled by the Renhua-Shaoguan Great Fault. The western peaks of Mount Danxiashan, centred around Bazhai, are 200 m higher than the rocks on the east bank of the Jinjiang River. All seven peaks in Mount Danxiashan with altitudes exceeding 500 m are found here.

Bazhai, at 619.2 m above sea level, is the highest peak in the area. It is followed by Yanyan at 594 m, Pingtou Village at 576 m, Chahufeng (Teapot Peak) at 556 m, Mount Guanyinshan at 517 m, and Luotuo Peak at 513 m. Peaks like Chuanyan, Wuxianyan, Shangtianlong, and Sister Peaks each have their unique characteristics.

The towering rocks, the Jinjiang River Valley, and various inter-mountain valleys of different sizes complement each other, showcasing the majestic scenery of Mount Danxiashan. Facing Bazhai is Teapot Peak, known as the 'No. 1 Pot Mountain in the World'. Teapot Peak resembles a giant purple clay teapot and is one of Mount Danxiashan's iconic landscapes. The journey typically takes 5–6 h.

3.10 Route 10—Jinjiang River Trail

The Left Bank Trail of Jinjiangg River stretches 3.5 km from Yangyuan Bridge to Changlaofeng Wharf. Along the way,

Fig. 7 Bazhai Peak, the highest point in Mount Danxiashan, reaches an elevation of 619.2 m. Its adjacent breathtaking landmark of Chahufeng (Teapot) Peak is a significant and captivating symbol of the region, exemplifying the distinctive landforms of the Danxia landscape

you'll pass through Yaotang Village, Danxiashan Museum (Visitor Center), the lower station of the cableway, and Zhanglaofeng Wharf.

Marvel at the picturesque scenery, including the landscape formed by the Sleeping Beauty, Elephants Coming Out of the Mountain, Parasol Rock, Jinyan Red Cliff, Mawei Waterfall, Hutouyan Fangshengtan Cliff, and the Jingui (Golden Turtle) Pilgrimage.

Throughout the four seasons, the trail showcases a variety of flowers and plants such as Danxia Wutong (Danxia Phoenix Tree), Danxia Xiaohua Jucao (Danxia Small-flowered Gesneriad) and Danxia Jincai (Danxia Violet). The pilgrimage cliff of the Jingui (Golden Turtle) is also a concentrated breeding ground for thousands of night herons and other birds. The journey takes approximately 2 h.

3.11 Shaoshi Scenic Area

Located adjacent to the eastern Zhenjiang River, the Shaoshi Scenic Area is steeped in legend and history. According to lore, Emperor Shun visited this area over 4,000 years ago, playing the Shaoyue, an ancient Chinese musical instrument, which led to its naming.

The area is comprised of thirty-six distinct peaks, including Shuangque Rock, Flagpole Rock, Chaoshiding, Jinguiyan, Mount Jinlong, Baizhai, and Wuyang Rock. Visitors can explore more than 40 sites, such as ancient fortress, cottages, temple ruins, and cliff coffins. It is also renowned as the original birthplace of Shao culture.

3.12 Feihuashui Scenic Spot

Located in the southwest of Mount Danxiashan, Feihuashui Scenic Spot comprises Matou Village, Feihuashui, Shilongxi Canyon, and the Kaixuanmen. Matou Village is the source of Feihua Water, where the Feihua Waterfall and Danxia Pot Caves combine perfectly. The Kaixuanmen is a natural arch that stands independently among the peaks. Shilongxi Canyon, the deepest canyon in Mount Danxiashan, is over 200 m deep, more than 4,000 m long, and varies from a few meters to over ten meters in width.

3.13 Zhenjiang Scenic Belt

The Zhenjiang River originates from Denglanzhai in Youshan Town, Xinfeng County, Jiangxi Province, and is part of the eastern source of the Beijiang water system in the Pearl River Basin. It has a drainage area of 7,554 km^2 and flows for 212 km. The last 24 km of its course flow through the southeast side of the scenic spot.

The Wumaguigou-Taiyangyan area, where the Zhenjiang River meets its tributary, the Jinjiang River, features the most dangerous currents, deepest canyons, most winding river courses, widest beaches, and most beautiful scenery in the entire Zhenjiang River Basin.

4 Ancient Fortresses: 'Hakka Wailou' (Enclosed Buildings) on Top of the Cliffs

Mount Danxiashan boasts the largest concentrated distribution area of danxia landforms in the low-latitude humid monsoon regions of China and even the world. The basin's altitude ranges from 50 to 619.2 m, and it features five levels of planation planes: the 600 m Bazhai plane, the 500 m Guanyinshan plane, the 400 m Jinlong Mountain plane, the 300 m Shuangque Rock plane, and the 200 m Shizishan Shizhai plane. The peaks, rock fortresses, and stone walls of the danxia landforms are deeply cut, highly discrete, and surrounded by cliffs, creating unique conditions for the construction of ancient cottages.

Mount Danxiashan was home to the Maba people 130,000 years ago and was later inhabited by the Baiyue people. Since the Qin (221–207 BC) and Han (202 BC–220 AD) Dynasties, many people from the Central Plains have ventured south of the Five Ridges to conquer and defend the border. Tang (618–907 AD) and Song (960–1276 AD) dynasties poet Zhiwen Song wrote in 'Early Development of Shaozhou': 'Zhuya Tianwai County, Tongzhu Hainan Standard'. During the Tang and Song Dynasties, Shaoguan's population increased significantly, leading to a rise in residents in Mount Danxiashan and its valleys.

Today, Mount Danxiashan covers an area of 292 km^2, encompassing 6 towns, 1 sub-district office, 14 administrative villages, and 76 natural villages in Renhua County and Zhenjiang District. The region has a population of 15,000, the majority of whom are Han Hakkas. The remaining Dan family, who lived by the water and made a living through fishing and hunting, along with the Lan and Zhong families scattered across the mountains, continue to tell the ancient and enduring history of the mountain people.

4.1 The Origin of the Fortresses

Northern Guangdong, southern Jiangxi, and western Fujian are the main areas where the Hakka people are concentrated, and they are also the primary distribution areas for the Hakka Wailou (enclosed, round, earth buildings). Since the Ming (1358–1644 AD) and Qing (1644–1911) Dynasties, thousands of these Hakka enclosing structures have been scattered throughout the Nanling and Wuyishan Mountains. During the Age of Cold Weapons, the Hakka Wailou served as fortified castles where entire Hakka villages and families lived and worked.

Constructing these walled buildings in the plains and river valleys required significant financial investment and could only be accomplished by pooling the manpower and material resources of an entire village and family.

In Mount Danxiashan, there are more than 100 ancient cottages, collectively known as the Danxiashan 108 Villages. These fortresses typically use natural defenses and are located near cliffs, which has led to them being called 'Hakka enclosures built on the cliffs'.

4.2 The Characteristics of the Fortresses

'Every mountain has a stronghold, and every stronghold has a gate, and every gate must be dangerous'. This saying reflects the significance of strongholds in Mount Danxiashan. The ancestors of the ancient Yue nationality in this region lived in natural caves among rocks. Mount Danxiashan is the area with the highest concentration of ancient cottages in Guangdong, many of which are built on steep, nearly inaccessible rocks. According to statistics, there are more than 100 ancient cottages throughout Mount Danxiashan, collectively known as the Danxiashan 108 Villages. These villages resemble castle-like hills surrounded by cliffs on all sides.

With more than 600 large and small, interesting rock outcrops and features in Mount Danxiashan, the names of villages far outnumber those of mountains. The fortress fulfilled multiple functions, such as providing refuge for villagers, serving as temples, operating as bandit strongholds, hosting military garrisons, and functioning as academies. A complete fortress often featured cliffs and natural barriers for protection, gates, and sheltering caves and pools inside the gates.

Features of the ancient fortress comprise:

- Natural Barriers and Gates: They were typically built on cliffs with elevations ranging from 200 to 600 m. These strongholds usually had more than two gates, such as the first gate, the second gate, or the third gate. The road to the mountain was rugged and dangerous, often consisting of single steppingstone steps. A single gatekeeper could defend against many attackers.
- Volume and Capacity: The fortresses had ample firewood, sufficient water sources, and food storage, providing basic living conditions.
- Terrain-Based Construction: Fortresses were built according to the terrain, utilizing caves or the gentle slopes of cliff tops for added safety.
- Local Building Materials: Rocks and stones were hewn for pools, and natural rock caves were widened for living spaces. Chiseled stones and cave sand were used for building walls and joining stones.

Fig. 8 Ximeizhai Fort is one of the most popular attractions of Danxiashan. It was built during Mid-to-late Ming Dynasty (~sixteenth century) as a shelter for locals fleeing late Ming conflicts (*Source* Hou R.F.)

- Water Supply: On narrow hilltops, water was precious. A complete fortress has pools and a water diversion system, and stone mortars were used for pounding rice.
- Famous fortresses: Zhanglao Village, Ba Village, Ximei Village, Tianyi Village, Ziwei Village, Pingtou Village, Tongjin Village, Jiucai Village, Bai Village, Jinlong Village, and Wuma Village are among the notable fortresses.

These ancient fortresses, with their high cliffs, intertwined vines, and weathered structures, provide a fascinating glimpse into the past. Over time, wind, rain, weathering, and plant roots have caused various levels of damage to many structures.

4.3 Introduction to Some of the Fortresses

The fortresses open to the public in Mount Danxiashan mainly include Zhanglao Village and Ximeizhai Village.

4.3.1 Zhanglao Village

Located in the main mountain of Mount Danxiashan, this village is a must-visit for tourists. Initially developed during the Tang and Song Dynasties, it flourished in the Ming Dynasty and reached its peak in the early Qing Dynasty. The cottage is divided into upper and lower parts, with the mountain gate of Biechuan Temple situated in the middle of the mountain and Xiaguan (also known as Haishan Gate) above the Danti iron lock natural barrier. In the Ming Dynasty, a mountain road connected it to Jinshiyan. In 1646, the Li Yongmao brothers bought the mountain, lived there, and later donated it to the Buddhist temple to build Biechuan Temple. Hundreds of monks and laypeople lived there, making it the liveliest cottage at that time. The mountain features horizontal caves of tens of square meters, including natural rocks, bamboo rocks, snow rocks, conch rocks, Wanxiu rocks, water curtain rocks, and schist rocks. Haishan Gate was demolished in 1999, the Mount Everest cable car was completed in 1998, and the Danxia plank road was finished in 2009 (Fig. 8).

4.3.2 *Ximeizhai Village* (in Mount Yangyuan)

This typical danxia rock wall is another must-see for tourists. The fortress was first built during the Chongzhen period of the Ming Dynasty, rebuilt in the ninth year of Xianfeng in the Qing Dynasty (1859), and repaired during the Republic of China. During the Anti-Japanese War, local people sought refuge here. One can reach the top of the village via the Jiujiu Sky Ladder and the Yunya Plank Road. The cliff inscriptions from the Qing Dynasty clearly record the history of Ximeizhai Village's establishment. There are hundreds of ancient fortresses in Mount Danxiashan that are not open to the public.

4.3.3 Sleeping Beauty Fortress Group

This area, the closest large-scale mountain block to Renhua County, contains six or seven fortresses. The most dangerous is Yangtou Village, with its intact gate and available living facilities such as pools, stone mills, and house foundations. On the southeast side of the mountain is Heshu Rock Village and the unfinished Luyi Hall, built during the Republic of China. Luyitang is one of the best-preserved fortresses in Mount Danxiashan. The cliffs near the fortress have developed small honeycomb-shaped caves larger than 30 m^2. Several Zhaimen houses are in the cliff cave on the north side.

4.3.4 Wentong Village

This general village contains the ruins of Wenfeng Academy, which are divided into upper and lower floors. The lower floor is large and exquisitely built, with a platform more than 100 m^2 wide and a pool with a capacity of over 20 m^3. The upper level is very hidden and can only be accessed with a rope ladder. The academy can accommodate dozens of people to study.

4.3.5 Bazhai

At an altitude of 619.2 m, Bazhai is the highest peak of Mount Danxiashan and a favourite among outdoor lovers. It is known for its supreme beauty. In the Qing Dynasty, Shitang Li's family purchased the mountain and built it. The mountain is shaped like a huge palm, with a cliff over 260 m high that was originally unclimbable. With the help of a gap in the cliff above the sky on the north side of Zhaimen, a 15 m high, nearly vertical 'ladder' was dug, allowing one person to go up and down. Above the ladder is the gate of the village, with shooting holes for defence. 'Dragon and phoenix' petroglyphs are engraved on the inner wall of the village gate. The multi-level horizontal caves in the village can accommodate hundreds or even thousands of people. The village is named after its owner, 'Yu Village'. Numerous reservoirs are dug at water drop points, and the top of the mountain is densely forested, with springs and wells that never dry up. The platform faces Teapot Peak, and the mountain features Slimming Stone, Shame Rock, and other scenic spots. One can reach the top by winding down the stone steps and enjoy the beautiful scenery of danxia landscape, such as the mitral peaks in the east.

4.3.6 Xinzhai in Tianyi Mountain

Built by the Ye family in Jiangtou Village during the Qianlong period of the Qing Dynasty, this large-scale cottage is close to the mountain. Dangerous areas are fortified, while calmer areas have walls and gates. The village contains dozens of houses, more than ten cliffs, and four large and small pools, with exquisite facilities and rigorous architecture. The roads are dangerous, the forest is deep, the spring is accessible, and firewood is available, making it an ideal refuge in the cold weapon era. Pingtou Village, located on the western peak of Mount Danxiashan, has the large and intricately carved Tianbiyan in the middle and lower part, and the dangerous Baimian Shenzhai in the middle and upper part.

4.3.7 Chuanyan Village

Located in the west of Mount Danxiashan, this village features a large natural hole on the top of the mountain. From Dajing Village to Zhaiding, there are four gates. Between Sanzhai Gate and Sizhai Gate are the ruins of Longhua Nunnery.

4.3.8 Huangchao Village

Situated on the west side of Mitama Peak, this village still has gun towers, gun holes, and defensive structures. Dozens of houses stand by the cliff, each about 10 m^2. Legend has it that the peasant uprising army of Huang Chao stationed here in the late Tang Dynasty.

4.3.9 *Tunjunzhai* (Sister Peaks)

At an altitude of 398 m, the east side is called Jinzhai, and the west side is known as Yunke Shanzhai. This village is close to the Jinjiang River, with cliffs on all sides and a vast mountain top. The inner alleys and valleys on the west side are well developed, known as the Danxia Labyrinth, and suitable for stationing troops. Several war horses are engraved on the cliff outside the gate of Yunke Shanzhai.

4.3.10 Jinlong Village

Located at the geographic center of Mount Danxiashan, Jinlong Mountain stands at 410 m above sea level. The mountains are dangerous, with dense forests. During troubled times, bandits often occupied the mountains and plundered passing ships. Before the construction of modern highways, travelers had to pass through Shaoguan City on foot via Wowo'ao, Aizhai Village, Jinjiang River, Fanguilao Village, Jinlongshan Rock Temple, and Jinguiyan Temple to reach Zhoutian and Nanxiong. Now, the cottages and mountain roads are deserted and difficult to traverse.

4.3.11 Wutai Shanzhai

Nestled among five mountains and rocks—Bazhai, Bashi, Yanyan, Bianzhai, and Xizhuyan—this village contains many temple ruins and rich cliffs. The most exquisite building is the Wutai Mountain Academy, located between the southeast cliffs of the mountain, which can accommodate 20 or 30 people studying on the cliff. Wuxianyan is a famous east-facing horizontal cave in Danxia Mountain. Sleeping in the rock, one can watch the sunrise of Danxia.

4.3.12 Yanyan Fortress Group

The second highest peak of Mount Danxiashan, Yanyan features majestic mountains, extensive horizontal caves, and lush vegetation. The northwest mountainside contains Yanyan Temple, while the middle and upper parts of the southeast house Xianqiaoyan Village, Guanglongyan Village, and Feiyunyan Village. Wild boars frequently roam the area.

5 Cliffs: History Carved into Rocks

As early as 129,000 years ago, the Maba people thrived in southern Guangdong, including the area of Mount Danxiashan. During the Neolithic Age, ancient Yue ancestors inhabited the region surrounding Mount Danxiashan. Over thousands of years, this area has fostered a rich Shao culture, also known as Danxia culture. By the Wei and Jin Dynasties, Shaoshi was already recognized as a famous mountain in China. The cliff inscriptions and petroglyphs on Mount Danxiashan highlight its rich historical and cultural heritage.

Unlike hard granite and limestone, which resist weathering, red sandstone is more susceptible to erosion. The oldest preserved cliff inscription in Mount Danxiashan is the 'Shejie Ji', carved on the cliff of Jinguiyan in the sixth year (1013 AD) of Dazhong Xiangfu during the Northern Song Dynasty. More than 300 inscriptions, dating from the Northern Song Dynasty to the Republic of China period, have been preserved in Mount Danxiashan. These include 12 from the Song Dynasty (960–1279 AD) and 10 from the Yuan Dynasty (1260–1271 AD) (Li 2013; Qiu 2013; Qiu and Li 2015). The cliff carvings of Mount Danxiashan are recognized as national key cultural relic protection units. Guangdong Province also boasts several cultural relic protection units, such as the Nyuzhuan Ruins, Xueyan Temple, Shiruquan, Ximei Village, Shiziyan Temple Ruins, and Mount Danxiashan Pagoda Tombs. Additionally, national level cultural relic protection units include the Jinguiyan Temple Ruins, Chuanlongyan Temple Ruins, and Jinjiang rock paintings, among others (Fig. 9).

Fig. 9 Ancient rock carving and painting could be found nearly everywhere on Mount Danxiashan and there are excellent historical and cultural records of interaction between rocks and humans

5.1 Jinjiang Petroglyphs

There are three notable sites for Jinjiang petroglyphs: Tangxia rock paintings, Baishi rock paintings, and the petroglyphs of Xiaba on the opposite bank of Jinshiyan (Hou 2019).

5.1.1 Tangxia Rock Paintings

Located on the left bank of the Jinjiang River, these rock paintings are downstream and close to the water surface, making them susceptible to submersion during flood periods. Five paintings are inscribed on the homogeneous sandstone wall, all carved in intaglio. The content includes characters, animals, poems, and inscriptions. Among them is a woman with a high crown, wide sleeves, and hands on hips, measuring 0.7 m high and 1.1 m wide. At the bottom right of the woman is a portrait of an official, standing 1.2 m high and 0.6 m wide. The scene also includes a cow and a dog, with the words 'Tianxia Taiping! Tianxia Taiping! Tianxia Taiping! engraved around the frame. These petroglyphs were carved in the second year of Tiansheng (1024 AD) during the Northern Song Dynasty (Li 2013; Qiu 2013).

5.1.2 Baishi Rock Paintings

Found on the right bank of the Jinjiang River, about 10 m above the river level, there are 23 petroglyphs in total. These are neatly carved on a red sandstone wall, approximately 27 m long and 1.6 m high. The lines are smooth and clear, depicting a variety of themes, such as people, deer, birds, fish, frogs, flowers, boats, buildings, the sun and moon, inscriptions, and abstract symbols. Notably, there are pictures of Sutra buildings, lotus and precious sons, and more than a dozen patterns of deer. These carvings date back to the sixth year of Baoyou in the Southern Song Dynasty (1258 AD).

5.1.3 Xiaba Petroglyphs

Situated on the right bank of the Jinjiang River, these petroglyphs include depictions of people and animals. However, the exact age of these depictions is unknown.

5.2 Shaoshi Cliff Group

Located in the Shaoshi Scenic Area, the Shaoshi Cliff Group includes notable sites such as Jinguiyan, Yongquanyan, Chuanlongyan Temple Site, and Jinlong Mountain. The history of this group spans thousands of years, from the Northern Song Dynasty to the Republic of China.

Among the cliffs, Jinguiyan is the oldest, featuring the 'Shejie Ji' inscription from the sixth year (1013 AD) of Dazhong Xiangfu in the Northern Song Dynasty. In the fourth year (1114 AD) of Zhenghe in the Northern Song Dynasty, 'The Story of the Golden Turtle Cliff' recorded the construction process of the Jinguishi (Golden Turtle Rock) Temple (Qiu and Li 2015).

The Chuanlongyan Temple site is notable for its large scale, with spacious and empty caves. The most captivating features are the east–west piercing holes developed along the soft rock layer.

5.3 Jinyan Cliff Inscription Group

Located in Jinshiyan, the Jinyan Cliff Inscription Group comprises over 80 cliff inscriptions. These inscriptions extend from Mengjueguan to Longwangyan, with the highest concentration found at Longwangyan. The inscriptions date from the fourth year of Xining in the Northern Song Dynasty (1071 AD) to the Republic of China period. The themes include inscriptions, titles, poems, travel notes, and more. Among these, eight Song engravings and eight Yuan engravings are considered the most valuable cultural relics. The 'Jinyan' cliff character is particularly notable, standing 1.8 m high and 1.4 m wide. It is carved on the north side of Longwangyan and was written by Zhao Rulei in the Southern Song Dynasty during the reign of Chunyou Bingwu (1246 AD).

5.4 The Cliff Group of Biechuansi Temple

Located both inside and outside the mountain gate of the Zen temple, the Cliff Group of Biechuansi Temple is divided into two parts by the old mountain gate: the 'Bieyoutian' cliff group inside the temple and the 'Danxia' cliff group outside the temple gate. Most of the carvings date from the early Qing Dynasty to the Republic of China period, with a few works from the Ming Dynasty.

The 'Danxia' cliff, situated outside the temple gate, is 2.7 m high, 2 m wide, and covers an area of 16 km^2. In the late Ming Dynasty, Li Chongmao, the master of Mount Danxiashan, wrote 'The Story of Mount Danxiashan.' This precious document, consisting of 39 lines with 35 characters each (totalling 1,368 characters), provides valuable insights into the history of Mount Danxiashan. Later, three poems from his brother Li Yongmao, titled 'Recalling Mount Danxiashan Dwelling,' were also engraved.

Inside the temple, there are notable inscriptions such as the 'Tianran Nunnery' and 'Fairy Trail' from the Ming Dynasty, and 'Kangxi Jiazi' (1684 AD) from the Qing Dynasty. The 'Bieyoutian' cliff is the largest in Mount Danxiashan, measuring 10.9 m in length and 3.6 m in height. The engraving is 0.5 m deep and covers an area of about 40 m^2.

5.5 Tongtian Gorge Cliff Group

The Tongtian Gorge Cliff Group is primarily distributed on both sides of the Danti Iron Cable and Yufeng Pavilion. Notable examples include inscriptions like 'Breathing Through the Sky' and 'Yiruo Ascending to the Sky,' which vividly describe the perilous nature of Danti.

5.6 Hailou (Conch) Peak Cliff Group

The Conch Peak Cliff Group features notable inscriptions such as 'Snow Rock', 'Conch Rock', and 'Most Superior'. 'Snow Rock', engraved in the reign of Kangxi Dingchou during the Qing Dynasty (1697 AD), is particularly impressive. Written in running script, it measures 2.48 m in height and 7.01 m in width. Even after more than three hundred years, the cyan mineral pigment on the surface of the font remains as clear as new.

5.7 Woodcut Calligraphy of Biechuansi Temple

The woodcut calligraphy preserved in the past dynasties of Biechuansi Temple is of great reference value for the study of the history of Mount Danxiashan. Han Shan Deqing, a prominent monk from the Ming Dynasty (1368–1644 AD), remarked on the challenges of life. In the early Qing Dynasty, an authentic couplet survived the 'Ban Xing Tang Ji' literary prison case in 1775: The wind passes through the bamboo forest; one can still see the temple. The cloud expands, concealing the clear water in the mountain.

6 Fairy House: The Literati Support the Country

Since the Wei and Jin Dynasties (220–589 AD), literati, poets, dignitaries, monks, and Taoists of all eras have travelled to and paused at Mount Danxiashan, leaving behind thousands of poems in praise. Selected parts are briefly introduced as follows (Fig. 10).

Fig. 10 A picture of Zhang Jiuling in around 720 AD (*Source* Baidu Pictures)

6.1 Zhang Jiuling (678–740 AD)

He was a renowned prime minister of the Tang Dynasty and a native of Qujiang, Shaozhou (now Shaoguan, Guangdong), is famous for his 'Two Poems of Senyu', which opens the widely acclaimed 'Three Hundred Poems of the Tang Dynasty'. His line, 'In this world and the next, this moment is forever', has remained timeless. His ode to Danxia Mountain is exemplified by 'The First Song in the River and Stream'.

6.2 Han Yu (768–824 AD)

He was one of the Eight Great Masters of the Tang and Song Dynasties, left behind famous lines such as 'The mountains and rivers of the Qujiang River have been renowned for a long time' and 'I wish to moor my boat at Shaoshi for a while'.

6.3 Hu Zeng (840–? AD)

A poet from the late Tang Dynasty (618–907 AD), is celebrated for his evocative verses. Among his renowned lines are: 'You know the desperate situation in the south of Lingnan', and 'The rain in Xili and the rest of the mountains is beautiful, and the smoke in the reeds exhausts, and the island looks blue'.

6.4 Yu Jing (1000–1064 AD)

A prominent minister of the Northern Song Dynasty and a native of Qujiang, Shaozhou (modern-day Shaoguan, Guangdong), was the first local literati to fully appreciate the danxia landscape, leaving behind a significant collection of poetry. Among his renowned works is 'You Shaoshi', a five-character rhythmic poem comprising 28 couplets and 280 characters in total, which stands as a representative piece celebrating the Danxia scenery. 'Shao Ting Ji' is acclaimed as the first beautiful essay describing the Danxia landform, while 'Renhua Jinshiyan' is recognized as the earliest poem dedicated to the Jinshiyan.

6.5 Su Shi (also Known as Su Dongpo) (1037–1101 AD)

He was a renowned writer of the Song Dynasty, celebrating the beauty of the Shaoshi area in his poem 'Wangshao Stone at Xiaodeng Jinshan Pavilion in Jianfeng Temple'. He praised the site with the lines, 'This place must be the house of immortals', capturing the enchanting scenery comprehensively.

6.6 Li Yongmao (1601–1648 AD)

He was a notable figure of the late Ming Dynasty and owner of Mount Danxiashan, who was born in Dengzhou, Henan. In 1646, Li Yongmao and his brother Li Chongmao invested over a hundred taels to buy the rocks around Changlaofeng to escape chaos. They named Zhanglaofeng 'Mount Danxiashan', after Mount Danxiashan in Nanyang, Henan, their hometown. Li Chongmao's essay, 'The Story of Danxia Mountain', spans 1,368 characters and details the efforts of Li Yongmao and other Ming Dynasty survivors in managing Mount Danxiashan. This work is considered a key resource for understanding the history and culture of Mount Danxiashan.

6.7 Li Chongmao and Others (1661 AD)

In the 18th year of the Shunzhi Emperor's reign during the Qing Dynasty (1661 AD), *Li Chongmao* established Mount Danxiashan as a dojo. The following year, Shi Dangui (1614–1680), a disciple of nature, arrived at Mount Danxiashan. During the subsequent five years, the Biechuansi Temple was constructed, reflecting the Buddha's principle of 'do not establish writing, teach beyond the scriptures'. Dan Guiying invited his teacher, Tianran Hanshi (1608–1685 AD), to become the first abbot of Biechuansi Temple.

6.8 Liao Yan (1644–1705 AD)

He was a native civilian literato, contributed over 20 poems and essays that vividly describe Mount Danxiashan in his collection, 'Twenty-seven Songtang Collection'. Notable works like 'You Mount Danxiashan' and 'Su Danxia' offer a documentary-style depiction of the danxia landscape during the early Qing Dynasty, painting a detailed and evocative picture of the region's beauty.

6.9 Yuan Mei (1716–1797 AD)

He was a prominent writer of the Qing Dynasty, composed 'You Danxia Ji' in the forty-ninth year of Qianlong's reign (1784). This work describes the Danxia landscape and its features.

7 Religion: Zen Beyond the Religion

A local saying advises, 'Choose Nanhua if desired, but opt for Danxia if otherwise inclined'. 'Nanhua' means Nanhua Temple, and 'Danxia' means Mount Danxiashan.

Mount Danxiashan and Nanhua Temple are intrinsically linked. Mount Danxiashan is a renowned landmark in Lingnan (South China) culture, and religion, especially Buddhism, plays a vital role. Over 20 ancient temples and temple ruins are scattered across the mountain. Among them, the temples and the religion history are of equal importance.

7.1 Jinshiyan Temple

Jinshiyan is one of the most geologically and landform-rich areas of Mount Danxiashan, also known for being one of the earliest developed cave groups on the mountain. Monks inhabited Jinshiyan during the late Tang and Five Dynasties periods. Fayun, a monk from the Northern Song Dynasty, travelled from afar and undertook large-scale construction here, establishing it as the earliest scenic spot in Renhua. Through various dynasties, it has seen periods of abandonment. Today, Jinshiyan Temple serves as a site for bhikkhuni (Buddhish nun) practice (Fig. 11).

7.2 Biechuansi Temple

Mount Danxiashan's rise in popularity since the early Qing Dynasty is closely tied to the construction and prosperity of Biechuan Temple. Founded in the first year of Kangxi (1662 AD), Biechuan Temple gained global recognition by recruiting many literati, scholars, and eminent monks who

Fig. 11 The historic Beichuan Temple, built in 1661 AD, is the heart of Mount Danxiashan which witnesses the latest development of this Redstone Park of China

turned to monastic life during the turbulent late Ming and early Qing Dynasties. Alongside Nanhua Temple and Yunmen Temple, Biechuansi Temple is one of the three major religious sites in northern Guangdong. Despite facing several abolishment, it remains a place for monks to practice today.

7.3 Mount Danxiashan Records

First published in the 38th year of Kangxi (1699) and updated in the 11th year of Yongzheng (1733 AD) over ten volumes, this record details the development of Mount Danxiashan from the first year of Kangxi's reign to the eleventh year of Yongzheng's reign.

7.4 Xianjuyan Taoist Temple

Located by Xianglonghu Lake, this temple is dedicated to Taoist practice.

8 The Six Weird Rocks of Mount Danxiashan

The extensive development of rock fortresses, walls, pillars, piercing holes, and natural bridges are defining features of the mature danxia landscape. From these diverse formations, people have selected a few with particularly unique shapes, dubbing them the 'Six Strange Stones of Mount Danxiashan'. These peculiar rocks showcase the remarkable and varied geological formations that make Mount Danxiashan a site of great natural wonder. They include rock pillars such as Yangyuanshi, Tianzhushi, Wangfushi, and Guanyinshi, and caves such as Yinyuanshi (formed along joints) and Longlinshi (formed by grouping of honeycombs).

8.1 Yangyuanshi (Male Rock)

A pinnacle that naturally grew and developed on the east side of Ximeizhai (Yangyuanshan) was discovered and named in 1992. It was named for its resemblance to male reproductive organs. The pillar is 28 m tall with a 7 m diameter and estimated weight of 600 tons, looks the most similar when viewed from east to west. It is the originator of many so-called 'Yangyuanshi' at home and abroad. The sun is shining in the morning, and the pillars are full of yang energy, which is the most suitable for viewing and worshiping. Known as 'the world's weirdest rock'.

8.2 Yinyuanshi (Female Rock) Stone

This interesting rock feature was discovered by villagers of the Xie Family in Qinghutang in 1998, was built as a viewing platform and opened to tourists. Located at the end of Xianglonghu Lake. It is a natural stone cave that resembles female reproductive organs. The Yinyuanshi is 10.3 m tall, 4.8 m wide and 4.3 m deep. The straight-line distance from Yangyuanshi is less than 3,000 m. Zhanglaofeng Peak resembles a huge screen separating the two rocks, and Jinjiang is like a silk belt connecting the two rocks.

8.3 Longlinshi (Dragon Scale) Rock

The Dragon Scale Rock is located on the inner wall of the Dragon King Cave, the last cave in Jinshiyan. Thousands of small honeycomb-shaped caves are connected in series like a dragon attached to the cliff wall. The dragon scales fluctuate with the seasons. The epiphytic microorganisms on the surface of dragon scales change color with changes in temperature and humidity, and they are called chameleons. It is light green in spring, dark green in summer, yellow green in autumn, and yellow brown in winter. Modern research suggests that the small honeycomb-shaped cave complexes known as dragon scales are the result of salt weathering. The relatively homogeneous and evenly developed thick sandstone layer with fine details, once exposed, will be attached to the cliff wall by the salt molecules in the air and water vapor. Over the years, the rock surface will be weathered differently, forming a large area of small honeycomb caves. The epiphytic microorganisms on the surface of the scales, known for their chameleon-like qualities, shift colors with variations in temperature and humidity. In spring, they are light green; in summer, dark green; in autumn, yellow green; and in winter, yellow brown. Modern research indicates that these small honeycomb caves, known as dragon scales, are formed by salt weathering. The relatively homogeneous and finely detailed thick sandstone layer, once exposed, becomes

Fig. 12 Tianzhushi or candle rock (*Source* Hou RF)

attached to the cliff wall by salt molecules in the air and water vapor. Over time, the rock surface weather is unevenly, creating a vast area of small honeycomb caves (Fig. 12).

8.4 Tianzhushi (or Candle) Rock

It is a pinnacle standing vertically in the landscape. The rock pillar has a relative height of 42 m and a diameter ranging from 5 to 8 m at its weathered base. And the base is seriously weathered. At the end of the Ming Dynasty, Li Yongmao described it in a poem 'Leave a single pillar to support the world and look down on the mountains and descendants.

8.5 Wangfushi (Watching Husband) Rock

It is a striking rock pillar attached to the cliff of Fengshuzhai. The formation bears a remarkable resemblance to a young, slender woman in a red dress, who appears to be gazing wistfully into the distance, longing for her husband's return.

According to local legend, her husband ventured out to fish, but due to the changing fortunes of life, the lake receded, and he has yet to come back. This poignant story has given Wangfushi its evocative name and has made it a symbol of enduring love and hope.

8.6 Guanyinshi (Goddess of Mercy) Rock

Also known as Sail Rock, Guanyinshi earned its name because it resembles the Goddess of Mercy, Guanyin, sitting gracefully on a lotus. Standing at an impressive 300 m, it is one of the tallest and most renowned rocks in Mount Danxiashan.This majestic rock formation depicts Guanyin holding a baby in her arms, with numerous smaller rocks surrounding her, resembling children seated and standing around her. Because of this unique appearance, it is also referred to as Songzi Guanyin (Guanyin with Children) and Boy Bai Guanyin. The most extraordinary feature of this rock is visible from behind, where the formation splits into three sections, resembling the petals of a lotus flower and giving it the appearance of a sail.

9 Three Major Waterfalls

Mount Danxiashan is home to hundreds of waterfalls, varying in size and seasonality. Among them, three stand out prominently:

9.1 Maweibao (Horsetail) Waterfall

Cascading from Jinyan Chipi (red cliff of Jinshiyan) at a height of 130 m, this waterfall flows through the upper part of the Jinshiyan section of the Danxia Formation and the upper part of the Bazhai section. It creates a perfect picture with Chibi Waterfall and Jinyan. The waterfall is most majestic after heavy rain, often resembling a horsetail, and evening rainbows are a common sight. However, it ceases to flow during long droughts.

9.2 Jiuniankeng Double Waterfalls

Located southeast of Mount Danxiashan, the Jiuniankeng area boasts two waterfalls, one large and one small. Both are approximately 70 m high, with the larger waterfall flowing year-round and the smaller one active for about half the year. Rainbows frequently appear under the sun, creating a mesmerizing spectacle.

9.3 Feihuaxi Waterfalls

The Feihuaxi area features upper, middle, and lower waterfalls, each around 15 m in height. These waterfalls are accompanied by a series of potholes, creating a breathtaking view. Below the waterfalls lies the 4-km long Shilongxi Canyon.

10 The Three Musketeers of Danxia Cliff Plants

Mount Danxiashan, with its expansive exposed cliffs, experiences a strong heat island effect and isolated island effect, making it an exceptional testing ground for the evolution of numerous life forms. One of the most remarkable features of the natural landscape is its rich and vibrant cliff vegetation. The plants are closely associated with history and culture in form of local identity and in practical usage, particularly for ornamental and medical reasons. Over 30 new plant species have been documented in Mount Danxiashan, the majority of which are cliff-dwellers.

Among these, the Danxia sycamore stands out as the representative species. Danxia sycamore, round-leaf small stones, and crape myrtle frequently grow together, forming the most eye-catching and drought-tolerant small trees and shrubs on the cliffs and edges of Mount Danxiashan. Despite their differing appearances, these plants are all brimming with character and are fondly referred to as the 'Danxia Three Musketeers'.

10.1 Danxia Sycamore

The Danxia sycamore, a national secondary protected plant, was discovered and named in 1986. Belonging to the Malvaceae family, this small deciduous tree follows a fascinating annual cycle:
February: Pulses rise.

March: Buds open.
April: Leaves grow.
May: Flowers bloom.
June: Leaves flourish.
July: Seeds set.
August: Fruits develop.
September: Leaves turn yellow.
October: Leaves fall.
November–January (Winter): Roots and lodges.

Throughout the year, the Danxia sycamore stands as a clear indicator of seasonal changes at Mount Danxiashan, with its bare branches lasting for over four months. It is both drought-tolerant and barren-resistant. With its purple flowers, follicles, and golden leaves, the Danxia sycamore is highly ornamental and has earned the nickname 'the biological business card of Guangdong'. The saying, 'Everything is safe in summer, and the sycamore is alone in autumn' highlights its unique presence throughout the seasons.

10.2 Small Stones with Round Leaves

Small stones with round leaves, known scientifically as Rosaceae Xiaoshiji, is an evergreen shrub. It is discontinuously distributed across Guangdong, Taiwan, and the Ryukyu Islands. In Guangdong, it is exclusively found on Mount Danxiashan. The leaves of this charming shrub are feather-like. Each March, small white flowers, delicate velvet blooms, and green fruits appear, adding a cute and appealing touch to the landscape. By September, the berry-like fruits mature into a striking sauce-purple colour.

10.3 Lagerstroemia Indica

Lagerstroemia indica, commonly known as crepe myrtle, is a deciduous and semi-deciduous small tree or shrub belonging to the Lagerstroemiaceae family. It is one of the indicator plants found on the top of Mount Danxiashan. Each year in midsummer, clusters of flowers resembling baskets bloom abundantly across the hills, creating a stunning display. The flowering period extends impressively for as long as a hundred days.

11 The Three Treasures of Mount Danxiashan

As the saying goes, Danxia boasts three treasures: red beans, orchids, and reviving grass.

11.1 Red Beans

In the Tang Dynasty poet Wang Wei's famous lines, 'Red beans grow in the southern country, and a few branches will grow in spring. May you pick more, this thing will be the most lovesick', the red beans he refers to are those found in the Mount Danxiashan area of the southern country. The Fabaceae red bean is an evergreen tree. Among the four varieties of red beans found on Mount Danxiashan, the most common is the soft pod red bean. The red beans here are vivid red, clear, and small in size. Additionally, the seed sprouts of red beans are slightly poisonous, which means they do not attract insects and remain intact for a long time. These qualities make red beans perfect symbols of enduring love between couples, as they can be preserved and cherished forever. Patches of red bean forests are found on major hills of Mount Danxiashan, including Yangzhou Village and Ba Village. For those seeking to collect these symbolic red beans, Conch Peak is a popular destination.

11.2 Orchids

The Danxia orchid, also known as the Shaoguan city flower, is represented by more than 50 varieties on Mount Danxiashan, including Chunlan, Qiulan, Suxinlan, Leafless Orchid, and Dendrobium Orchid. These orchids are widely clustered in the jungle or on the cliffs at the top of Mount Danxiashan. Dendrobium orchids often grow symbiotically with Shixiantao on the cliffs. A notable addition to this diverse flora is Danxialan, a new variety of the Orchidaceae Leafless Orchid family, named in 2013. This species is mainly distributed along the deep sandy valleys and occasionally grows on cliffs. Danxia orchid is characterized by its individual leaves, slender needle-like stems, and a plant height of less than three centimeters. Its small marsh orchids, which are difficult to distinguish, inspire awe at the marvels of creation. It is generally referring to Chinese orchid.

11.3 Resurrection Grass (Selaginella)

Danxia Mountain is home to 11 known species of the Selaginellaceae family. Renowned as one of the world's most drought-tolerant plants, Selaginellaceae can survive up to three years of dormancy and revive when exposed to water. These resilient plants are widely distributed across the exposed rock walls of Mount Danxiashan, making them a fascinating sight for visitors. In 2021, a new species of Selaginellaceae, Oriental Selaginella, was named after Mount Danxiashan as the type specimen, highlighting the mountain's unique botanical diversity.

Mount Danxiashan, named after the danxia landscape, stands out globally for its exceptional biodiversity and rich cultural heritage. It proudly holds the title of the number one sacred mountain in China's danxia region, a testament to its unparalleled significance.

12 Conclusions

Mount Danxiashan exemplifies the enduring and profound relationship between humanity and nature. Its geology and landscape foster the development of a unique ecological environment, supporting special species, as well as religious and cultural growth. This Redstone Park of China showcases the traditional wisdom of local communities, their determination to overcome challenges, and their adaptability in achieving a better quality of life through sustainable use of available resources and the environment.

References

Ai TH, Zhang YS (1994) An explanation of the name origin of Mount Danxiashan. In: Proceedings of the second academic symposium on tourism development of Danxia landscape, economic geography (Supplement), pp 48–53

Chen GD, Liu HS (2008) Geology of the Gushui River Basin in Jiangxi. In: Complete works of Chen Guoda, vol 1. Central South University Press, Changsha

Danxiashan (2024) Hilking and driving routes of Danxiashan, series 1–10 (Pamphlets in Chinese). Danxiashan Administrative Committee, Danxiashan UNESCO Global Geopark, Shaoguan, China

Fang Q et al (2022) Guangdong Danxia Mountain plant illustrated book. Science Press, Beijing

Guo FS et al (2022) Chatting about Danxia. Science Press, Beijing. Shaoguan historical documents series. Guangdong People's Publishing House, Guangzhou

Hou RF (2019). Records of Mount Danxiashan. Collation notes, "Danxia" quarterly, vol 21, Renhua, Guangdong, China

Huang J (2009) Landforms of Mount Danxishan. Science Press, Beijing

Li CM (2013) Record of Mount Danxiashan. In: Jiang Q (ed) Collection of ancient cliff inscriptions of Mount Danxiashan. Xiling Seal Engravers' Society Publishing House, Hangzhou

Qiu J (ed) (2013) Ancient cliff inscription collection of Mount Danxiashan. Xiling Yinshe Publishing House, Hangzhou

Qiu J, Li F (eds) (2015). Mount Danxiashan records (of Qing Dynasty, in 1733), photocopied from the Shanghai Library. The Qing-era Renhua County Annals describe it as a "shelter for locals fleeing late Ming conflicts."

Management and Sustainable Development

Fang Chen and Li Yuqiang

Abstract

Mount Danxiashan has a rich history of human activity, marked by three phases. The first phase centered on basic livelihood practices like fishing, hunting, and farming. The second phase introduced religious and cultural pursuits such as temple-building, pilgrimages, and admiration of the mountain's landscapes. The third and current phase focuses on conservation, sustainable management, and integrated resource use.

1 Background

Mount Danxiashan has a rich history of human activity, marked by three phases. The first phase centered on basic livelihood practices like fishing, hunting, and farming. The second phase introduced religious and cultural pursuits such as temple-building, pilgrimages, and admiration of the mountain's landscapes. The third and current phase focuses on conservation, sustainable management, and integrated resource use (Fig. 1).

1.1 Early Stage

The earliest inhabitants of this red stone park used the fertile river valley for farming, fishing, and hunting, while living in natural caves. Evidence of human activity, dating back over 4,000 years to the Neolithic Age, includes numerous cultural relics and nearly 100 sites where homes were carved into rock caves. This distinctive practice shaped Danxiashan village culture, reflected in the saying, 'Every mountain has a village, every village has a gate, and every gate must be a stronghold'. Peaks like Bazhai, Changlaozhai, and Jiucai Village derive their names from these historic settlements.

Mount Danxiashan's human history spans three stages of utilization. The first involved basic livelihood practices like fishing, hunting, and farming. The second saw the emergence of religious and cultural activities such as temple-building, pilgrimages, and landscape admiration. The third and ongoing stage emphasizes modern conservation, sustainable management, and integrated resource use (Huang 1994; Danxiashan Administrative Committee 2011) (Fig. 2).

1.2 Growing Stage

The primary inhabitants of Mount Danxiashan have historically been the Dan and Hakka people. The Dan people, traditionally reliant on fishing and living on boats in the Jinjiang and Zhenjiang River basins, faced challenging conditions that limited their population. Post-1949, government efforts helped them transition to land-based living by providing housing and farmland. Today, fishing within the Danxiashan UNESCO Global Geopark is prohibited for ecological conservation, with limited operations allowed in designated areas of the outer Zhenjiang River Basin.

Despite its humid subtropical climate and freshwater resources, Mount Danxiashan's rugged terrain and nutrient-poor red sandstone soils hinder large-scale agriculture. The Hakka people adapted by settling in fertile mountain basins, growing fruit trees and crops on slopes, and cultivating rice and vegetables in plains. Known for their strong clan structures and emphasis on education, Hakka communities integrate Buddhist, Taoist, and Confucian values into daily life,

F. Chen (✉)
Danxiashan Administrative Committee, Danxiashan UNESCO Global Geopark, Shaoguan City, Guangdong Province, China
e-mail: 398475377@qq.com

L. Yuqiang
Beijing Zosen International Engineering Consulting Co. Ltd., Beijing, China

Y. Ng and W. Eder (eds.), *The Redstone Park of China*, Geoheritage, Geoparks and Geotourism,
https://doi.org/10.1007/978-3-031-95603-4_7

Fig. 1 An ancient settlement made use of the natural caves in Mount Danxiashan

Fig. 2 4,000 years ago, Mount Danxiashan was known as 'Shao Stone', located in what is now the southeastern region of the geopark

Fig. 3 Rural countryside in Mount Danxiashan

promoting sustainability. Ownership of surrounding mountains and fields has been well-preserved, fostering a harmonious rural landscape of mountains, rivers, and villages.

For centuries, Mount Danxiashan's unique features have attracted religious practitioners, particularly Buddhists and Taoists. Monks and priests have meditated in its caves since the Sui and Tang Dynasties, reaching peak activity during the Ming and Qing Dynasties. Temples like Jinshiyan and Biezhuan remain significant, surrounded by red sandstone cliffs adorned with carvings, statues, and paintings. These artifacts document human interaction with the mountain, including ownership records of natural resources. Rich in stone tablets and ancient trees, these areas contribute to both historical preservation and conservation (Figs. 3 and 4).

1.3 Modern Stage

Since the founding of the People's Republic of China, Danxiashan's conservation and resource management have been prioritized, initiating government-led efforts in sustainable utilization. In 1963, the Renhua County Government established the Mount Danxiashan Forest Farm, followed by the Mount Danxiashan Management Office in 1976 to strengthen conservation. The opening of the Elder Peak to tourism in 1980 led to the creation of the Mount Danxiashan Scenic Management Bureau, alongside enhanced protection measures like a police station and a joint forest protection team with village committees.

Designated as a national key scenic spot in 1988, Danxiashan became a national nature reserve in 1995, spanning 292 km^2. The Renhua County Government set protection boundaries and signed agreements with local villages to assign forest protection responsibilities, supported by financial compensation. In the same year, management regulations for the scenic area and nature reserve were revised to improve local policies.

Recognized as a national geopark in 2001, Danxiashan gained global geopark status in 2004, marking its international significance. Its management was transferred to the Shaoguan Municipal Government in 2006, resulting in a department-level agency and standardized management practices, despite periodic reforms in subsequent years (Ng 2011). Mount Danxiashan was listed as one of the 49 national park candidate areas by the central government in 2022 and will be jointly managed by the central, provincial, and municipal governments in the future (National Forestry and Grassland Administration et al. 2022) (Fig. 5).

2 Historical Changes in the Management Structure

The management of Mount Danxiashan has evolved through three distinct stages based on three principles:

- Limited Use with Respect for Nature
- Voluntary Protection under Township Regulations and Civil Agreements
- Protection and Management in Accordance with Laws and Regulations

This progression illustrates the increasing formalization and regulatory oversight of management practices over time.

Fig. 4 A special monument for Biezhuan Temple was set up by the Shaoguan Prefecture in the 31st year of Qianlong's reign (AD 1766) to prohibit intrusion into the Temple

2.1 The Stage of Respecting Nature and Limited Utilization

This stage was characterized by a deep respect for nature and traditional farming, shaped by ancient Chinese philosophical principles. Limited productivity and cultural traditions drove early Mount Danxiashan inhabitants to protect their environment, ensuring survival and ecosystem preservation.

Blending religious beliefs, Feng Shui principles, Confucian values, and Hakka clan traditions, environmental protection became a cultural norm. Proverbs such as 'Don't hunt March birds' and traditions like 'worshiping mountains and ancestors' reflect deliberate efforts towards conservation. Practical ideas, like cherishing food, were passed down and remain relevant today. These ecological concepts inspired communities to preserve their environment and use resources sustainably. Spanning millennia, they continue to shape modern practices and serve as a foundation for nature conservation (Fig. 6).

Fig. 5 The 'Danxia on the Water' project was planned and built in the 1990s, is one of the most popular tourist destinations *(Source: Xie JS)*

2.2 The Stage of Spontaneous Protection Through Township Regulations and Civil Covenants

Over the past thousand years, the Hakka people sought refuge from wars and invasions, leading to a significant population increase in Mount Danxiashan and the rapid establishment of ancient villages such as: Zhangwu Ancient Village (Zhang clan and Zhang Jiuling); Xiafu Ancient Village (Li clan and Li Bai); Niubi Ancient Village (Zhu clan and Zhu Xi); Lingtou Lingbei Village (Zhou clan and Zhou Dunyi); other villages by the Wu, Xie, Ma, Liu, Qiu, and Ye clans.

To maintain harmony, these Hakka villages implemented township regulations and civil agreements to govern and protect natural resources such as mountains, forests, and water sources. These agreements, based on shared understanding, set boundaries, allocated resources, and established norms. Developed collectively, they formed an early legal framework rooted in community consensus and contractual principles.

Fig. 6 The historical cliff inscriptions of Mount Danxiashan reflect over a millennium of cultural and religious heritage

The resource rights defined in these agreements have persisted through the centuries and continue to underpin procedures for confirming and certifying resource ownership today.

2.3 Protection and Management Stage in Accordance with Laws and Regulations

Since the establishment of a management agency for Mount Danxiashan, its administration has progressed through three stages:

- 1963–2006: Led by Renhua County, focusing on forest resource protection.
- 2006–2009: Managed by Shaoguan City, emphasizing tourism development.
- 2009–Present: A joint effort by Guangdong Province and Shaoguan City, balancing protection and development.

The management structure evolved from prioritizing forest conservation to integrating tourism and, eventually, adopting a sustainable approach. In 2023, preparations began to establish Mount Danxiashan National Park under a new national park framework, marking the start of joint governance by central, provincial, and municipal governments. The current approach focuses on protecting scenic resources while planning and operating tourist attractions (Table 1).

In summary, the evolution of Mount Danxiashan's management structure has progressively enhanced resource protection. What began with a primary focus on forest conservation expanded to include the promotion of geotourism and has now advanced to a holistic approach that safeguards the danxia landforms and their surrounding ecosystem (Peng et al. 2005a, b; Chen et al. 2015). At every stage, the core principles of conservation have been upheld, ensuring the preservation of valuable resources to support the sustainable development of Mount Danxiashan (Fig. 7).

3 Current Management System

Since 2016, the management of Mount Danxiashan has followed a collaborative model, with co-management by provincial and municipal authorities. Shaoguan takes the lead, coordinating efforts at the county and district levels. The framework distinguishes between government and enterprise roles, as well as management and business functions. Administrative rights over resources such as forests, rivers, land, and cultural relics are shared by the Renhua County and Zhenjiang District governments. Village collectives manage fields, orchards, and forest land near villages, while residential properties are privately

Table 1 Changes in Danxia Mountain protection and management organizations from 1963 to 2023

Stage	Year	Organisation	Establishment by	Management content	Management area km^2
1	1963	Mount Danxiashan Forest Farm	Renhua County Government	Protect the forest resources of Mount Danxiashan (Zhanglaofeng (Elder Peak) area)	30
2	1976	Mount Danxiashan Management Office	Renhua County Government	Protect and manage the forest and scenic resources in the Zhanglaofeng (Elder's Peak) area of Mount Danxiashan	30
3	1980	Mount Danxiashan Scenic Management Bureau (Mount Danxiashan China Travel Service)/Main Division level	Renhua County Government	Protect and manage Mount Danxiashan scenic resources, plan, construct and operate Mount Danxiashan tourist attractions	30
4	1994	Mount Danxiashan Scenic Management Bureau (Deputy Division level)	Renhua County Government	Protect and manage the scenic resources of the Mount Danxiashan National Scenic Area, plan, construct, manage and operate the Mount Danxiashan Tourist Area, and manage the three administrative villages in the scenic area	215
5	2006	Shaoguan City Mount Danxiashan Scenic Area Management Committee/ Directorate level	Shaoguan Municipal Government	Responsible for planning, development, management and protection of Mount Danxiashan Scenic Area. It's Mount Danxiashan Tourism Investment Company looking after the operation of the scenic area	292
6	2009	Guangdong Shaoguan Mount Danxiashan National Nature Reserve Administration/ Directorate level	Guangdong Provincial Government	Responsible for the specific management and protection of Mount Danxiashan National Nature Reserve	280
7	2009	Shaoguan Mount Danxiashan Scenic Area Management Committee (Shaoguan Mount Danxiashan Global Geopark Management Committee) and Guangdong Shaoguan Mount Danxiashan National Nature Reserve Management Bureau/Director-Level	Shaoguan Municipal Government	To protect and manage the Mount Danxiashan Scenic Area, nature reserve, cultural and natural heritage, and geopark; to promote and implement relevant laws and regulations; to draft and implement plans; to protect and utilize natural and cultural resources; to review and supervise land use and construction projects; to manage religious sites and cultural relics; and to establish rules to ensure the normal operation of tourism	292

(continued)

Table 1 (continued)

Stage	Year	Organisation	Establishment by	Management content	Management area km^2
8	2012	Shaoguan Mount Danxiashan Management Committee and Guangdong Shaoguan Danxia Mountain National Nature Reserve Management Bureau/Shaoguan City Mount Danxiashan Tourism Industry Park Management Committee/Director-Level	Shaoguan Municipal Government	To protect and manage the Mount Danxiashan Scenic Area, nature reserve, cultural and natural heritage, and geopark; to promote and implement relevant laws and regulations; to draft and implement plans; to manage the scenic area's land and forestry resources, religious sites, and cultural relics; to oversee environmental protection; to construct and maintain infrastructure; to conduct comprehensive law enforcement within the scenic area; and to oversee the administration of 14 administrative villages and the Mount Danxiashan Tourism Investment and Management Company	1. 292 (Danxia Mountain Nature Reserve Area) 2. 300 (Planning and construction work for the Mount Danxiashan Eco-Industry Park)
9	2016	Shaoguan Mount Danxiashan Management Committee and Guangdong Shaoguan Danxia Mountain National Nature Reserve Management Bureau/Director-Level	Shaoguan Municipal Government	To be responsible for the protection and industry management of the scenic area, as well as the review and approval of construction projects within the scenic area and the Mount Danxiashan tourism industry park. To exercise supervisory and management authority over human activities and resource protection within the area, and, if necessary, to entrust relevant departments or institutions in Renhua County and Zhenjiang District with administrative enforcement powers	1. 292 (Mount Danxiashan Nature Reserve Area) 2. 300 (Preliminary Review of Construction Projects in the Mount Danxiashan Eco-Industry Park)
10	2023	To establish Mount Danxiashan National Park and to prepare for the establishment of the Mount Danxiashan National Park Management Bureau			

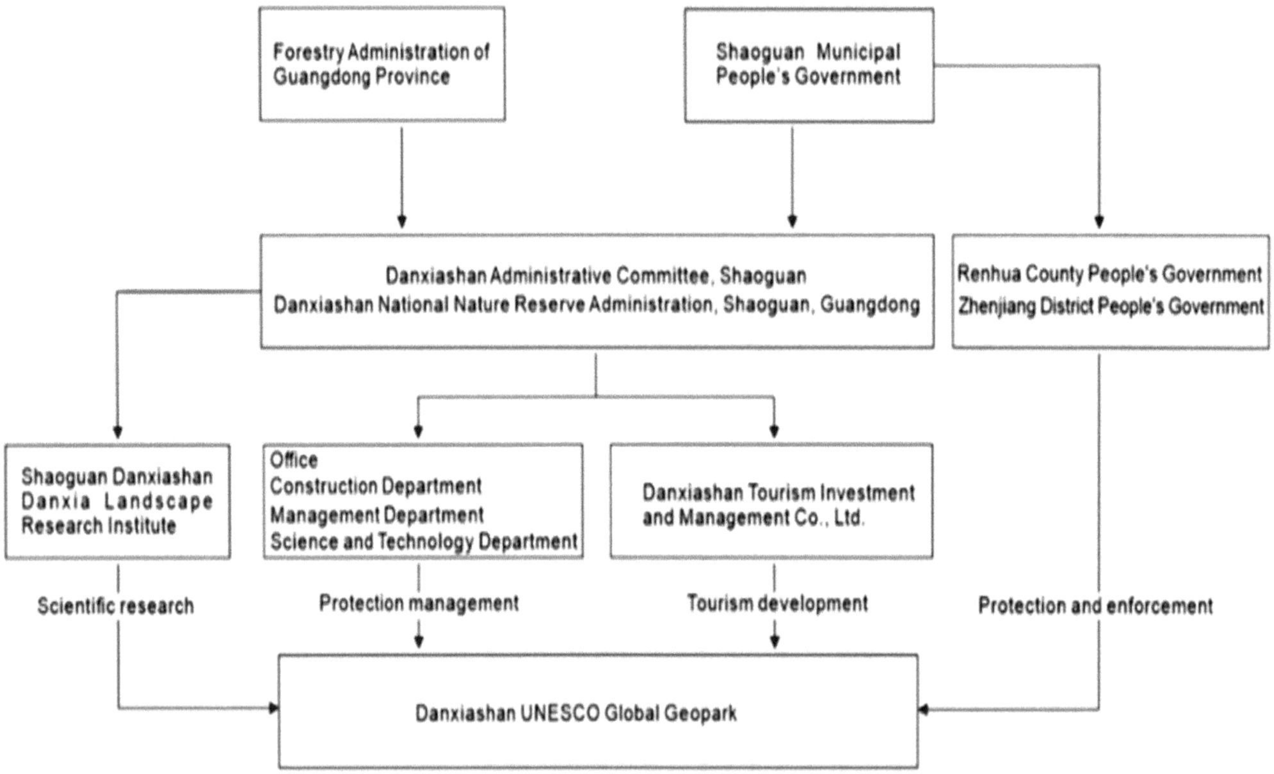

Fig. 7 Position of Danxiashan UNESCO Global Geopark at the provincial and municipal administrative hierarchy

owned. Service facilities are owned by enterprises, with clearly defined boundaries. Guided by principles of 'scientific planning, unified management, strict protection, and sustainable utilization', this approach has consistently protected resources, improved management practices, and fostered stable growth in tourism, creating a model that integrates conservation and development.

3.1 Current Management Structure

The Shaoguan Mount Danxiashan Management Committee, in collaboration with the Guangdong Shaoguan Mount Danxiashan National Nature Reserve Management Bureau, is the governing body for Mount Danxiashan. This director-level institution under the Shaoguan Municipal Government consists of four departments: Office Administration, Construction, Management, and Technology.

The committee oversees the protection, utilization, and management of Mount Danxiashan, reviews construction projects in the tourism area, and coordinates with industry departments. Local governments handle law enforcement for resources, planning, forestry, environmental protection, and tourism. The committee reports to the Shaoguan Municipal Government and receives guidance from national and provincial forestry administrations.

Spanning Renhua County and Zhenjiang District, Mount Danxiashan's management includes key local government leaders. Clearly defined job target agreements outline roles and objectives, ensuring effective governance. The organization employs 26 full-time staff (15 in management, 3 logistics staff, and 8 conservation workers) and 5 part-time staff (Fig. 8).

3.2 Management Effectiveness

3.2.1 Protection

(1) Implementation of 'One Park, One Law' Protective measures adhere to national and provincial laws like the Forest Law and Wildlife Protection Law. In 2009, the Guangdong Provincial Government issued the Regulations on the Protection and Management of Mount Danxiashan, while work on the Protection Ordinance is ongoing, providing a legal framework under the principle of 'One Park, One Law'.

(2) Comprehensive Planning System Plans such as the Shaoguan Mount Danxiashan Research and Education Development Plan and the Scenic Area Plan (2011, 2025) designate 30% of the park as a no-development zone, 55% as a strictly controlled research zone, and 15% as a general

Fig. 8 Management structure of within the Danxiashan UNESCO Global Geopark in 2024 (*Source* Danxiashan Management Administration).

control zone for eco-tourism. Only 10% of the park is open to tourism, prioritizing the protection of geological relics and ecosystems.

(3) Comprehensive Resource Protection

- All forest lands are classified as ecological public welfare forests and insured; a 2016 program redeemed 1,600 ha of artificial forests for 1.39 billion yuan (USD 0.195 billion).
- Geological documentation has identified 387 relic sites, 88 of significant value. Hazard monitoring points and updated signage enhance protection.
- Surveys identified 2,048 wild plant species and 2,036 animal species, including 30 nationally protected plants, 56 protected animals, and 39 new species like Danxia orchid.
- Cultural surveys registered 165 relics, including temples and cliff carvings. A database of 99 ancient sites, published cliff inscriptions, temple repairs, and plans for post road protection highlight preservation efforts.
- The 'River Director' system monitors water quality in the Zhenjiang and Jinjiang rivers, maintaining surface water quality at or above Class II standards (Fig. 9 and Table 2).

3.2.2 Management System

Mount Danxiashan employs a comprehensive management system, including work responsibility frameworks, staff performance regulations, duty schedules, protection patrols, forest fire prevention, and guide service management. Job target responsibility agreements signed by all staff clearly define duties and goals, while standardized regulations ensure efficient management.

The Management Committee works with Renhua County and Zhenjiang District through a joint meeting system. Patrol-identified issues are promptly addressed by law enforcement, ensuring compliance and protecting the reserve. The committee's key responsibilities include resource protection, utilization, and management, as well as reviewing construction projects and coordinating with industry departments. Local governments retain law enforcement roles related to land use, planning, and environmental protection.

Guided by principles of scientific planning, unified management, strict protection, and sustainable utilization, the committee maintains the park's authenticity and ecosystem integrity. Since 2010, the site has been rated excellent by the IUCN and has passed four UNESCO evaluations without major violations, incidents, or forest fires, underscoring its effective and internationally recognized conservation practices (Fig. 10).

3.2.3 Information Technology

In recent years, the Management Committee has launched the innovative Smart Mount Danxiashan project, integrating advanced technology into the park's management and protection efforts. Key initiatives include the development of a grid-based wildlife monitoring system, a real-time monitoring and management platform for the tourist area, an air quality monitoring station, and geological relic and disaster monitoring stations. A comprehensive database has also been established to catalog cultural relics survey results (Figs. 11 and 12).

Additionally, a 3D model of Mount Danxiashan has been created to enhance spatial analysis and planning. Forest fire monitoring equipment has been deployed, and drones are actively utilized for forest disaster prevention and control,

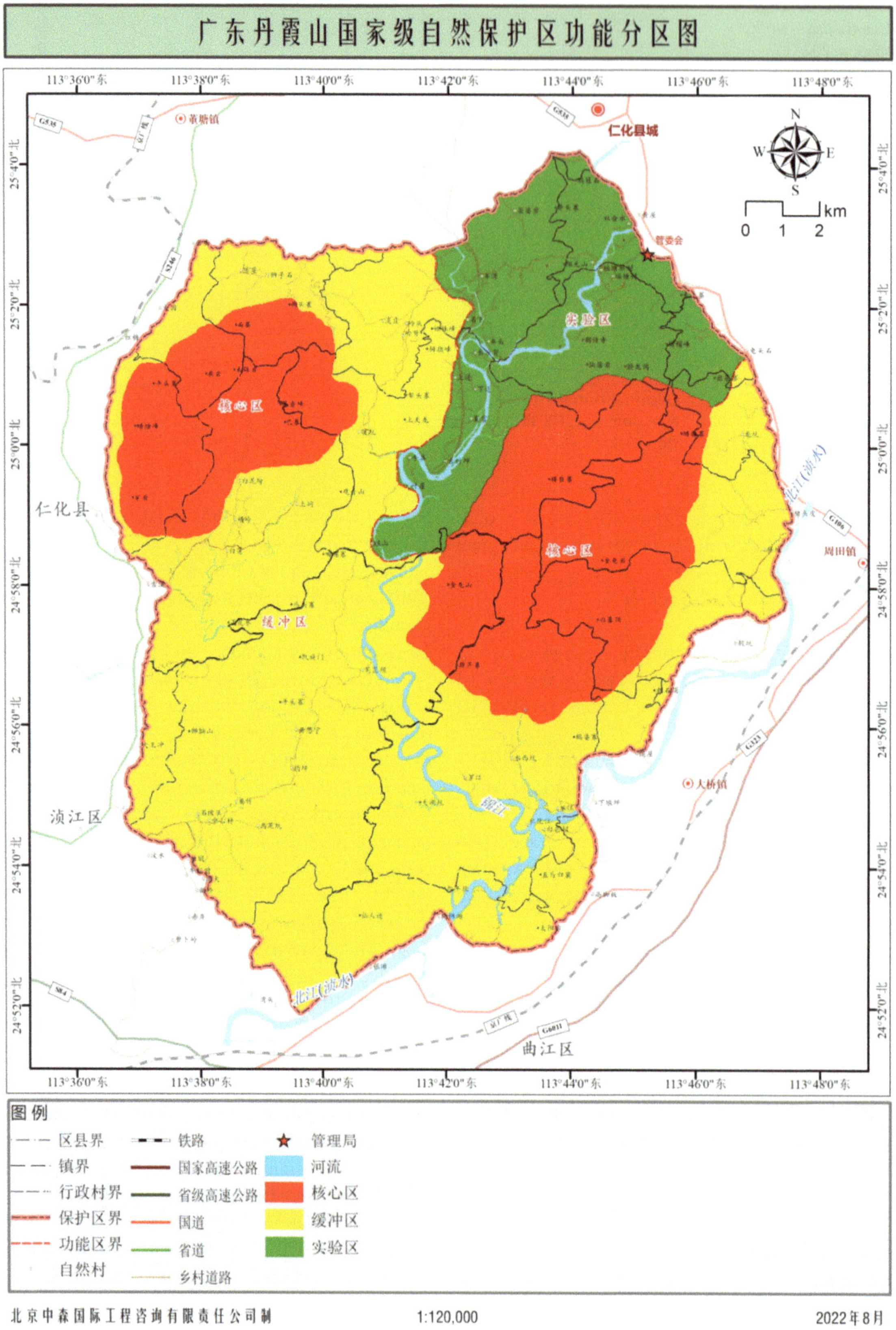

Fig. 9 Mount Danxiashan National Nature Reserve functional zoning map (Legend: **Red star**—Management Office; **Boxes:** Red—Core zone, Yellow—Buffer zone, Green—Experimental zone. **Lines**: Grey, shorter broken—County boundary, Grey, longer broken—Town boundary, red solid—Boundary of protected area, red broken—Functional boundary, black/white—Railway, brown—National Highway, dark green- Provincial Highway, thinner red—National Road, green—Provincial Road, khaki—Village Road)

Table 2 Comparison of protection effectiveness for conservation targets in Mount Danxiashan National Nature Reserve (Remarks: NPU = National Protection Unit, PPU = Provincial Protection Unit)

Types	Survey of 2010		Survey up to October 2022		Comparison	
	Quantity	National importance	Quantity	National importance	Changes in number	National importance
Fern	139	Class I: 1 Class II: 4	154	Class II: 6	15	Class I: 1 Class II: + 2
Seed plant	1567	Class I:0 Class II: 8	1733	Class I:1 Class II: 23	166	Class I + 1 Class II: + 2
Insects	1023	Class II:1	1517	Class II: 2	494	Class II: + 1
Fish	100	0	37	0	−63	0
Amphibians	21	Class II: 1	29	Class II: 1	8	0
Reptiles	41	Class I: 1 Class II: 5	72	Class II: 8	31	Class I: −1 Class II: + 3
Bird	157	Class II: 16	255	Class I: 2 Class II: 41	98	Class I: + 2 Class II: + 25
Mammal	88	Class I: 3 Class II: 11	48	Class II: 2	−40	Class I: −3 Class II: −9
Geoheritage	319	–	319	–	0	–
Cultural relics	84	NPU: 0 PPU: 2	165	NPU: 1 PPU: 5	81	NPU: + 1 PPU: + 3

Fig. 10 Immediate removal of potential falling rocks for public safety

ensuring a modernized and efficient approach to the park's conservation and safety.

3.2.4 Scientific Research and Public Education

Mount Danxiashan is a national leader in scientific research and public education, serving as a model for natural education, research, and science outreach. The Management Committee collaborates with over 30 domestic and international universities and research institutions, such as Sun Yat-sen University, to develop databases and monitoring systems for geological relics, hazards, and biodiversity. Over 200 experts have studied topics like geomorphology, biodiversity, archaeology, and community development, resulting in more than 1,000 published academic papers.

Founded in 2019, the Danxia Geomorphology Research Institute promotes research on Danxia landforms, hosting the world's first UNESCO Global Geopark training base. It fosters collaboration between World Heritage sites and geoparks, regularly participating in national and international conferences. In 2021, it launched a project documenting a century of danxia geomorphology research, preserving and advancing studies of these unique landforms.

On the educational front, Mount Danxiashan has developed a national research practice base for primary and secondary students, offering over 200 research courses and organizing competitions, exhibitions, and publications. Recognized for its excellence, it has been designated as a National Science Education Base and a National Forestry and Grassland Science Education Base, among other honors. The

Fig. 11 Infrared cameras captured a silver pheasant, a national second-class protected wild animal, with the image transmitted right back to control centre in real time

Fig. 12 Infrared cameras have captured images of a leopard cat, a second-class national protected wild animal, in its natural habitat

park also established the Mount Danxiashan Natural School and Guangdong's first Science Popularization Town.

3.2.5 Community Development

The Mount Danxiashan area includes 8 towns, 27 administrative villages, and 247 natural villages in Renhua County and Zhenjiang District, with 9 administrative villages, 47 natural villages, and 11,285 residents within the park. In recent years, governments have invested 962 million yuan (USD 135 million) in rural revitalization, building beautiful villages, enhancing tourism infrastructure, and integrating agriculture, culture, and tourism. Over 500 Danxia homestays and farmhouses have been established in the park's open area, engaging more than 10,000 residents in ecotourism activities.

A provincial citrus industry park surrounds the park, spanning 200,000 mu (13,333 hectares) and generating an output value of 1.5 billion yuan (USD 0.21 billion). It was the first in China to receive the title of 'China's Climate Ecological Premium Products' and features renowned local products like Changba Shatian pomelo, Danxia tribute orange, and Danxia white hair tea. Xiafu Village also became the site of the first national geological cultural village. This has fostered a coordinated development pattern of eco-industries in communities around Mount Danxiashan.

3.2.6 Tourism Development

Mount Danxiashan has maintained a focus on scientific development and strict protection, turning its lush landscapes into valuable assets. Since the 1980s, geology experts like Zeng Zhaoxuan, Huang Jin, Peng Hua, and Wu Zhenyang (Young Ng) have guided the planning of three scenic areas Zhanglaofeng Peak, Yangyuan Mountain-Tongtai Bridge, and Xianglong Lake-Wolonggang, as well as two sightseeing water projects on the Jinjiang River. Public facilities such as the Mount Danxiashan Museum, nine scientific exploration routes, and three nature education trails accommodate over 3 million visitors annually.

Designated a Global Geopark in 2004 and a World Natural Heritage site in 2010, Mount Danxiashan's international recognition has significantly boosted its visibility and visitor numbers. Since 2010, the site has attracted about 3 million visitors annually, generating 60 million yuan in ticket sales and 600 million yuan in overall tourism revenue (Weng and Peng 2014). Its success in science education has drawn more science and research tourists, particularly youth and families, making it a key contributor to community development. Tourism revenue supports both park management and the sustainable growth of surrounding communities, fostering a harmonious model of coordinated development.

3.2.7 Infrastructure Construction

In 2019, Mount Danxiashan established the South, Southwest, West, North, and Main Gates near key towns and village clusters, enhancing recreational opportunities and development. The 46-km Reading Danxia Road was created using fire prevention pathways and renovated rural roads. This scenic route, featuring seven integrated service centers in surrounding villages, forms a rural tourism corridor linking park entrances with local communities.

The initiative has drawn tourists to explore rural landscapes, boosting eco-tourism and agricultural product sales. It has also enabled residents to benefit directly from the park's conservation and sustainable development efforts.

4 Sustainable Development Model

Mount Danxiashan is a multi-brand composite nature reserve in China, recognized nationally and globally for its geological, biodiversity, and cultural significance. Its Danxia landforms, biodiversity, and cultural heritage have received designations from authorities like the State Council and UNESCO, with the park listed in two of UNESCO's three major international categories.

This branding system ensures the protection and sustainable use of Danxiashan's assets, supported by legal frameworks for National Key Scenic Areas, National Nature Reserves, and National Geoparks. Global Geopark and World Natural Heritage designations also contribute significantly to development, tourism, and education initiatives (Peng et al. 2005a, b).

In the past decade, eco-tourism activities such as research, science education, and practice have bolstered Danxiashan's reputation as a leading scientific and educational site. Its century-long legacy of Danxia landscape research has earned acclaim provincially, nationally, and internationally.

Managing this unique site requires scientific expertise, strong coordination, and global perspectives. Reforms have clarified protection responsibilities and enhanced administration. The park aligns comprehensive planning systems to ensure harmony between development and conservation while integrating international and domestic regulations. Collaborative governance and community engagement have enabled effective policies, increased funding, and coordinated brand development supported by diverse agencies (Figs. 13, 14, 15 and Tables 3, 4, 5, 6).

5 Management Challenges and Solutions

5.1 Management Challenges

5.1.1 Overlapping Geopark, Scenic Area, and Nature Reserve, Along with Zoning Challenges

In 1993, Mount Danxiashan applied for nature reserve status, but its proposed area overlaps significantly with the Mount Danxiashan National Scenic Area (approved in 1988), including villages, farmlands, and tourism zones. With only 15.28% of the area (in the northeast) designated as an experimental zone, development projects have stalled. Following the 2019 reform, unifying nature reserves under the State Forestry and Grassland Administration, advanced methods were used

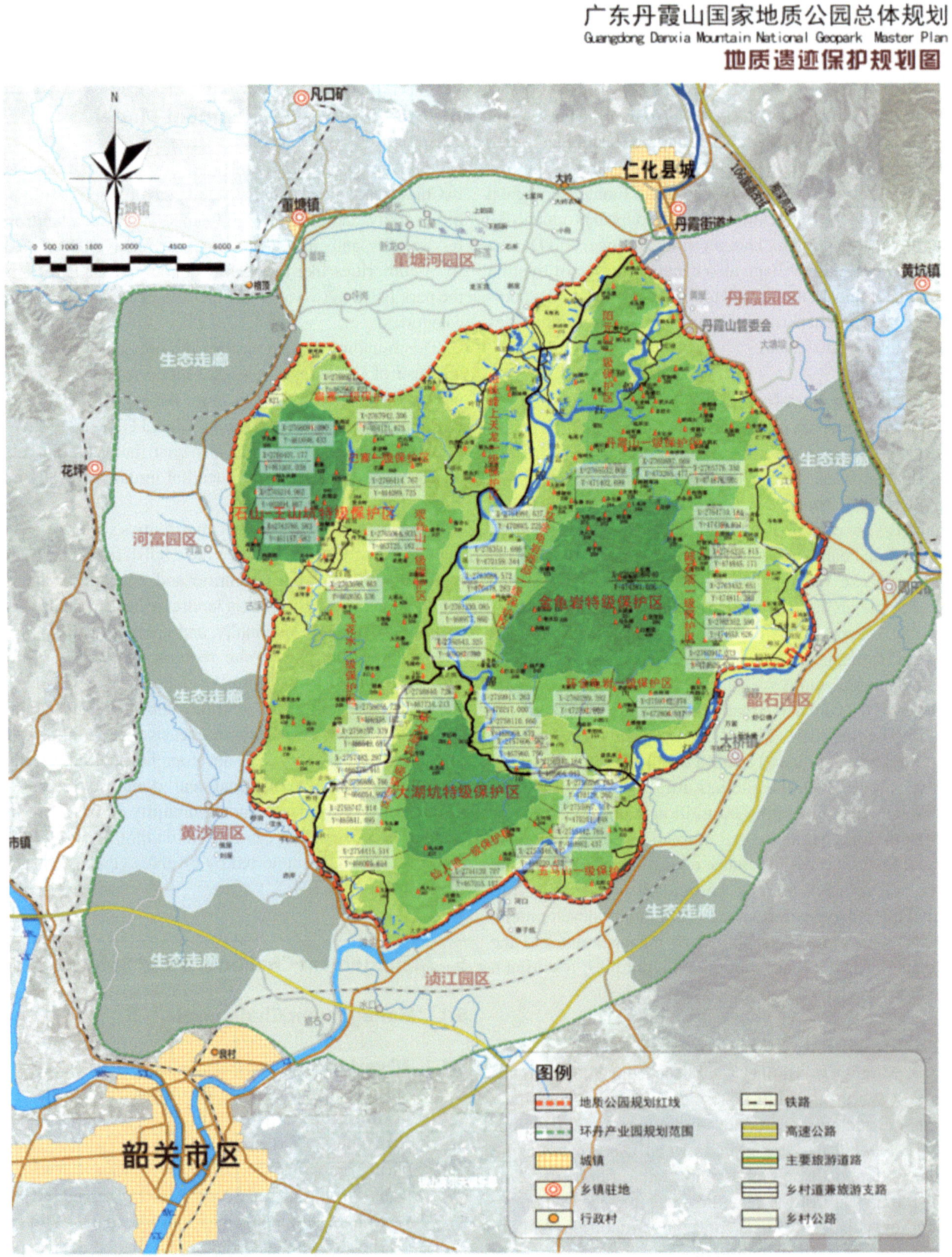

Fig. 13 Functional Map of Mount Danxiashan National Geopark (Red dot line- geopark boundary, the green areas are different levels of protection areas)

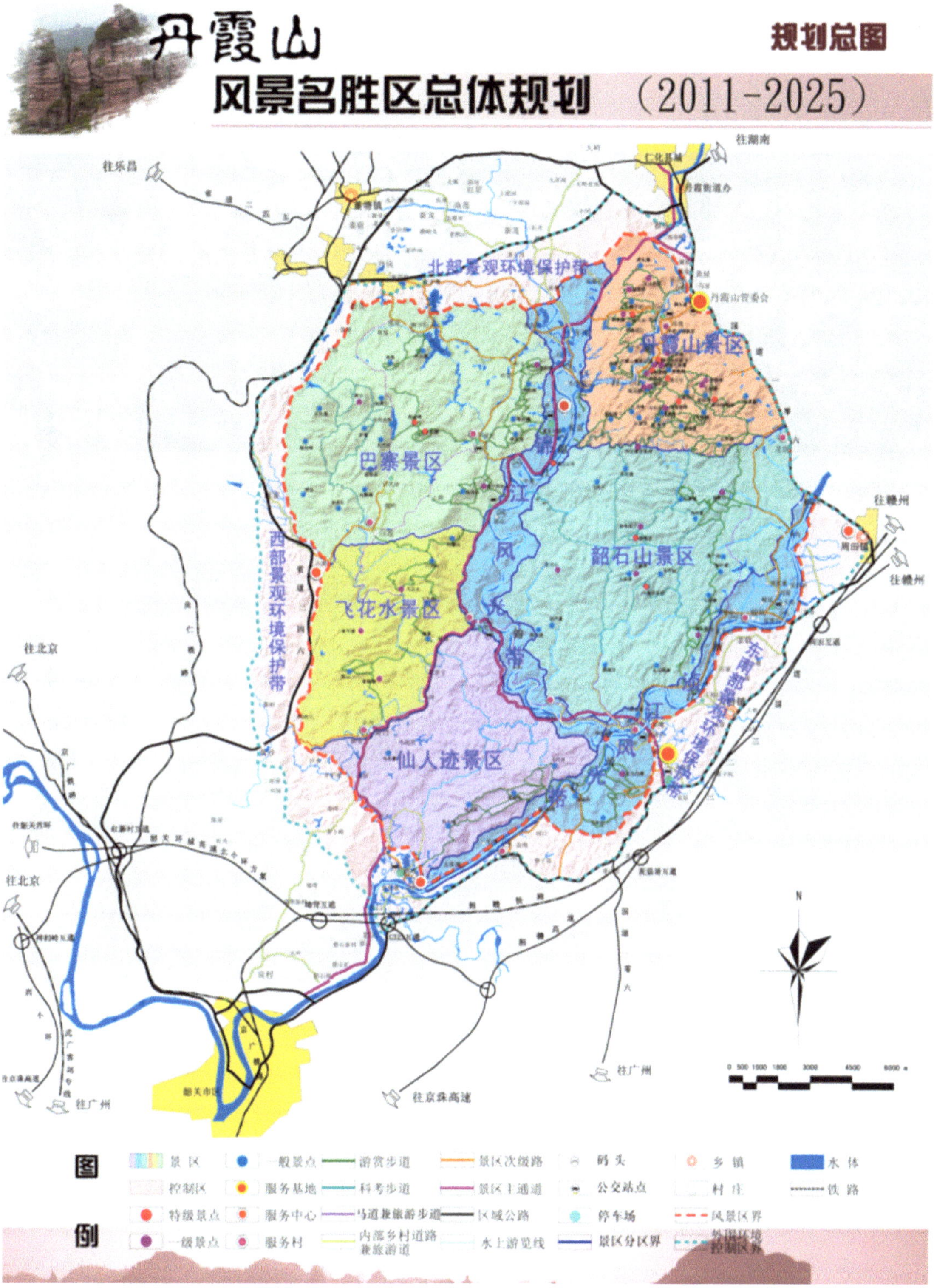

Fig. 14 Danxiashan National Scenic Area Planning Map showing the six scenic districts of the geopark within the red boundary

Fig. 15 Hakka villages established among the mountains and rivers have been here for thousands of years

Table 3 Summary of international brands of Mount Danxiashan

No	Brand	Protected area/Protected target/Management requirements	Organizing entity and schedule	Legal/Management base
1	Global Geopark	292 km^2 of danxia landform and geological landscape/Protect geological relics, disseminate Earth sciences, promote community sustainable development	UNESCO 2004	Global Geopark Construction Guidelines/ Evaluation System/Mount Danxiashan Geopark Plan'
2	World Heritage Site	168 km^2/danxia landform's earth scientific value, unique danxia landscape/ Protect, research, and utilize for the common benefits of humanity	UNESCO 2010	World Heritage Convention/World Heritage Periodic Monitoring Report/'China Danxia Protection and Management Plan'/ 'Guangdong Danxia Mountain Protection and Management Plan'
3	UNESCO Global Geopark Mentorship Training Base	Using Danxiashan Global Geopark as a base, park managers serve as mentors to help train foreign geopark management personnel	UNESCO 2017	Become a UNESCO partner and execute training tasks according to the commissioned training agreement

to delineate and zone the reserve, though final approval is pending.

5.1.2 Strengthening Regional Management and Monitoring System

Mount Danxiashan spans both Renhua County and Zhenjiang District. Historically, protection and tourism development have concentrated on the northeastern section in Renhua County, leaving the southern area in Zhenjiang District in need of enhanced management. Given the park's considerable size, extending 22.9 km from north to south, it is essential to establish additional management stations, maintenance points, and inspection posts to complete an effective patrol and management system.

Table 4 National Branding of Mount Danxiashan

Number	Brand	Protected area/Protected target/Management requirements	Organizing entity and schedule	Legal/Management base
1	National Scenic Area	292 km^2/Danxia landform natural landscape and Mount Danxiashan/ Suitable for tourism or conducting scientific and cultural activities	State Council 1988	'Regulations on Scenic Areas'/'Guangdong Mount Danxiashan Scenic Area Overall Plan'
2		280 km^2/Concentrated distribution area of Danxia landform geological relics/ Preserve natural baseline, reserve species, scientific research, and education base	State Council 1995	Regulations on National Nature Reserves/ Comprehensive Plan for Guangdong Mount Danxiashan National Nature Reserve
3	National 5A-Level Tourist Attraction	Open Scenic Area/The highest grade of tourist attractions in China, representing world-class boutique tourist attractions/ Evaluated based on 12 criteria including tourism transportation, safety, sanitation, resources, and visitor satisfaction sampling, balancing social and economic benefits	National Tourism Administration 2000/2012	'Tourism Law'/ 'Classification and Evaluation of Quality Grades of Tourist Attractions'/'Mount Danxiashan Tourism Concept Plan'
4	National Geopark	292 km^2/A concentrated distribution area consisting of Danxia landform geological relics, natural geomorphological landscapes, and cultural landscapes/Protect geological relics, promote science, popularize knowledge, and use tourism resources to develop the economy and serve society	Ministry of Land and Resources 2001	'Measures for the Construction and Management of National Geoparks'/'Overall Plan for Guangdong Mount Danxiashan National Geopark'
6	National Land and Resources Science Education Base	Open scenic areas and park villages/Showcase and promote the protection, research, and utilization achievements of the danxia landform	Ministry of Land and Resources 2012	'Science Popularization Law'/'Management Measures for National Land and Resources Science Education Base
7	Member of the China Man and the Biosphere (MAB) Network	292 km^2/Ecosystem of the danxia landform geographical area/Integrate scientific research, environmental monitoring, talent training, demonstration roles, and local community participation to protect ecosystem and biological genetic diversity	China MAB Secretariat 2013	Convention on Biological Diversity

(continued)

Table 4 (continued)

Number	Brand	Protected area/Protected target/Management requirements	Organizing entity and schedule	Legal/Management base
8	National Key Cultural Relics Protection Unit	Zhanglaofeng Peak Scenic Area/111 Cliff Inscriptions from the Song, Yuan, Ming, and Qing Dynasties at Mount Danxiashan/The highest protection level for immovable cultural relics with significant historical, artistic, and scientific value, essential for studying the evolution of human civilization and considered a common heritage of humanity	2013 State Council 2013	Cultural Relics Law'/ 'Danxiashan Cliff Inscriptions Protection Plan
9	National Science Education Base	Open scenic areas and park villages/Utilize Mount Danxiashan's science popularization resources to conduct science and education activities for the public, enhancing public scientific and cultural literacy	China Association for Science and Technology 2015	'Law of the People's Republic of China on Science Popularization'/ 'National Science Education Base Standards'
10	National Primary and Secondary School Research Practice Base	Open scenic areas and park villages/Organize and conduct research practice educational activities to enhance the social responsibility, innovative spirit, and practical ability of primary and secondary school students	Ministry of Education 2017	Science Popularization Law/Compulsory Education Law
11	National Long-Term Forestry and Grassland Research Base	292 km^2/Improve the construction level of research bases, fully utilize the fundamental and strategic role of research bases	National Forestry and Grassland Administration 2019	Management Measures for National Long-Term Forestry and Grassland Research Bases'
12	National Nature Education School	Open scenic areas and park villages/Organize and conduct nature education activities with significant results and a demonstrative effect	China Forestry Society 2019	
13	National Forestry and Grassland Science Education Base	Nature Reserve Experimental Area National Forestry and Grassland Administration/	Ministry of Science and Technology/2023	

5.1.3 Strengthening Ecosystem Research

Mount Danxiashan has been a leader in geological and geomorphological research both domestically and internationally. Recently, it enhanced wildlife monitoring with a 24/7 infrared network. However, continuous research on forest ecosystem processes and functions remains insufficient. Greater investment is needed in research, strengthened partnerships with domestic and international institutions, and the establishment of specialized research centers and systems. Tailored scientific monitoring plans should be developed, integrating high-tech methods to study ecosystems and key species, while advancing long-term monitoring of the danxia landform ecosystem.

Table 5 Provincial branding of Mount Danxiashan

Number	Brand	Organizing entity and schedule	Management requirements
1	Guangdong Province Land and Resources Science Education Base	2012 Guangdong Provincial Department of Natural Resources	Conduct natural resources science education activities
2	Guangdong Province Environmental Education Base	2015/Guangdong Provincial Department of Ecology and Environment	Conduct environmental protection science education activities
3	Mount Danxiashan Nature School	2015/Guangdong Provincial Department of Ecology and Environment	Conduct nature education activities
4	Guangdong Province Primary and Secondary School Research Practice Base	2017/Guangdong Provincial Department of Education	Serving primary and secondary school research practice
5	Guangdong Province Key Cultural Relics Protection Unit	2017–2022/Guangdong Provincial Cultural Relics Bureau	Protect and revitalize the use of cultural relics
6	Guangdong Province Nature Education Base	2019/Guangdong Provincial Forestry Bureau	Provide ecological nature education services
7	Guangdong Province Research Travel Base	2019/Guangdong Provincial Department of Culture and Tourism	Provide research travel products and services
8	Guangdong Province Science Popularization Town	2019/Guangdong Provincial Department of Science and Technology, Guangdong Provincial Association for Science and Technology, Guangdong Provincial Department of Culture and Tourism Leverage	Mount Danxiashan scientific research advantages, cultivate a learning community, and promote comprehensive and participatory development and governance
9	Guangdong Province High-Quality Nature Education Base	2022/Guangdong Provincial Forestry Bureau	Demonstrate and drive the transformation of nature reserves to provide high-quality nature education services
10	Guangdong Province University Student Practice Teaching Base	2022/Guangdong Provincial Department of Education	Serving university student teaching internships and practice
11	Guangdong Province Environmental Education Base Demonstration Unit	2022/Guangdong Provincial Department of Ecology and Environment	Demonstrate and lead in providing excellent environmental education
12	Top Ten Science Education Bases in Guangdong Province	2022/Guangdong Provincial Association for Science and Technology	Demonstrate and drive science education

5.1.4 Collectively Owned Forest Land, Forest Quality Needs Improvement

The forest land in Mount Danxiashan is mostly collectively owned. Although all forest land is included in public welfare forest management, it encompasses about 10% of artificial collective commercial forest, leading to severe restrictions on the forest management rights of village collectives and villagers. Currently, continuous patches of Masson pine in Mount Danxiashan are frequently affected by pine wood nematode disease, which is difficult to control.

5.1.5 Personnel Structure Needs Optimization

The Management Committee has a relatively small number of staff, with an average age of about 50 years. There is a lack of essential professionals in geographic information systems, geology, wildlife protection, ecology, and tourism management, resulting in insufficient vitality.

5.1.6 Slow Development and Weak Tourism Market

Since 1980, tourism has been confined to the 10 km^2 Danxia Scenic Area, representing just 3% of the park. Functional zoning restrictions have stalled new tourism projects for 20 years, resulting in a single tourism format and slow industry

Table 6 Comparison of the overall plans of the three major national nature reserves of Mount Danxiashan brands

Type	National key scenic area	National nature reserve	National geopark
Key Elements	Natural and cultural landscapes dominated by danxia landform	Danxia landform natural landscape and natural ecosystem	Geological relics and danxia landform
Nature and Positioning	A national park with overall landscape features dominated by danxia landform, as well as religious culture and historical sites	Typical geological and geomorphological phenomena and natural ecosystem reserves	A natural park of typical geological relics dominated by danxia landform
Main Functions Key	Protection and enjoyment of natural and cultural landscapes	Protection of various natural landscapes, ecosystems, and biodiversity	Geological relic protection and geological science education tourism
Content	Protection of natural and cultural landscapes Landscape and sightseeing planning and design Tourism development and attraction construction Tourism facility planning and design	Ecological environment protection Biodiversity protection Scientific research and science education tourism Protection facilities	Geological relic protection Ecological environment protection Scientific research and science education tourism Sustainable community development
Planning Characteristics	Emphasize the protection of various landscapes and the environment Focus on the cultivation of the scenic value of various landscapes Emphasize the development of sightseeing tourism	Emphasize natural and ecological protection Focus on ecological and science education tourism Emphasize geological scientific research	Emphasize the protection of geological and geomorphological relics Focus on geological science education tourism and research Emphasize the development of scientific exploration and science education tourism
Relationship between Development and Protection	Strict protection, sustainable use	Protection-oriented, limited use, no tourism development in the core and buffer zones	Balance between protection and development
Approving authorities	State Council of the People's Republic of China	State Council of the People's Republic of China	Ministry of Natural Resources

growth. Peak tourist seasons are holidays and weekends, with visitors mainly from the Guangdong-Hong Kong-Macao region for sightseeing. Weekday and low-season potential remains untapped. The state-owned Danxiashan Tourism Investment Operation Co., Ltd. manages the area, but tourism products are updated slowly, the enterprise lacks vitality, there is a shortage of skilled professionals, and operational capacity needs improvement.

5.1.7 Uneven Regional Development and Slow Growth in Remote Villages

The Mount Danxiashan area spans 2 counties, 7 towns (subdistricts), 9 administrative villages, and 47 natural villages, with a population of 11,285. While some Danxia Subdistrict villagers are involved in tourism services, most rely on traditional farming. Protection regulations in buffer zone villages restrict development, leaving communities stagnant. Limited job opportunities force many young and middle-aged residents to work elsewhere, resulting in significant rural hollowing.

5.2 Promoting High Quality Development Through National Park Establishment

5.2.1 National Policy Opportunities to Address Development Constraints

Since 2019, the Management Committee has utilized favorable policies to advance the construction of Mount Danxiashan National Park, resolving zoning issues in the nature reserve. By redefining control zones, they integrated rural and ecological tourism areas into the general control zone, paving the way for growth. In 2023, the park was named one of 49 candidate national parks under development following the State Council's layout plan.

5.2.2 Scientific Planning to Address Shortcomings

In 2019, a comprehensive plan was launched to guide Mount Danxiashan's development over the next decade, covering protection, ecological restoration, public education, sustainable resource use, disaster prevention, and smart management initiatives. The plan integrates the Yue Dan Highway, Jinjiang Green Waterway, and Zhenjiang Scenic Corridor with the

Danxia Science Popularization Zone and Xiafu Recreation Zone. A reservation-based, low-impact tour model emphasizes the 'in-park experience, out-of-park service' approach. Development of ten entrance communities is prioritized to strengthen eco-tourism facilities and foster coordinated local economic growth.

5.2.3 Establishing a 'Scientific and Smart Danxia'

Leveraging rich ecological resources, the park aims to build a strong research platform and collaborate with leading institutions to address gaps in expertise. Public education will be enhanced, empowering communities through science initiatives while fostering harmony between people and nature. Comprehensive surveys of flora and fauna will establish Mount Danxiashan as a "scientific mountain." Advanced monitoring technologies will develop a smart management platform to ensure effective governance.

5.2.4 'Five-Dimensional Integration' Heritage Education

Using a "five-dimensional integration" model, the Management Committee drives sustainable development by creating a popular science industry with community participation. Research is transformed into accessible science services, and residents are trained as science popularization experts. Homestays and farms are converted into education centres forming the Science Popularization Town. This initiative has attracted schools, families, and research institutions, boosting the park's reputation. The shift to guided tours also involves residents, enabling neighbouring areas to develop nature education and eco-tourism while diversifying employment opportunities and fostering sustainable growth.

6 Implications of Mount Danxiashan Management Model

6.1 Innovative Management Model—Community Co-Building, Governing, and Sharing

6.1.1 Establishing a Science Popularization Town

In 2019, Guangdong approved its first Science Popularization Town: Mount Danxiashan Science Popularization Town. This initiative integrates resources like the Mount Danxiashan Global Geopark, Nanling Meteorological Center, and Fankou National Mining Park under the Mount Danxiashan Management Authority. Thirty science classrooms with dedicated facilities, mentors, and curricula have been created to support a vibrant community of researchers, educators, and enthusiasts. The program encourages university graduates to launch businesses and helps locals transition to the science popularization industry. Families are inspired to establish science classrooms, enhancing knowledge, networks, and incomes while preserving heritage and sharing the Danxia story.

6.1.2 Encouraging Research and Science Education

Mount Danxiashan's management has clarified park resources through scientific research, technology adoption, and multi-level monitoring teams. Collaborating with over 30 universities and research institutes, such as Sun Yat-sen University, it advances research in Danxia geomorphology, biodiversity, and the environment. Expert teams provide intellectual support for science education. New technologies, including automated monitoring of air quality, geological relics, and wildlife, enhance real-time research and conservation efforts. These tools also supply updated, vivid materials for science education, showcasing the biodiversity value of Danxia Mountain while supporting ecological preservation.

6.1.3 Resource Integration for Community Science Education

(1) Enhancing science education

Leveraging its resources, Mount Danxiashan Reserve has developed nine themed science education routes, installed over 1,000 illustrated interpretation signs, and established birdwatching, butterfly watching, and firefly trails. Facilities like the biodiversity hall and classrooms in the Mount Danxiashan Museum have been upgraded to serve diverse audiences, providing a strong foundation for science education.

(2) Multi-dimensional science educator team

The Reserve has created a 'three-in-one' science educator team of over 600 members, including consultants, research mentors, and volunteers, through initiatives like science volunteer training camps. Collaborating with university experts, the Reserve has hosted lectures and training to aid residents in transitioning to the science education field. Prominent educators, such as 'Butterfly Sister' and 'Birdman,' have emerged, showcasing local expertise.

(3) Developing science education products

The Reserve has developed over 200 science education experiences, covering geology, astronomy, biodiversity, ecology, Hakka culture, and farming. Popular products, including

'Treading Danxia and Gazing at the Stars' and 'Making Floral Rice Wine,' have been well-received, appealing to a wide range of audiences.

(4) Branded science education activities

Mount Danxiashan has collaborated with agencies, universities, and residents to launch engaging initiatives like the Science Volunteer Training Camp, immersive education courses, and regional festivals. In 2020–2021, Science Town hosted nearly 200 public education activities and 200 nationwide lectures and exhibitions, enriching tourism with educational elements and enhancing community tourism development.

6.1.4 Rural Revitalization Through Science Education

Mount Danxiashan has become a science education and research hub, recognized nationally and regionally as a primary education base and as a national and provincial science education site in the Guangdong-Hong Kong-Macao Greater Bay Area. Since 2019, the Science Town initiative has drawn conservation managers from Thailand, Australia, Russia, and over 20 Chinese provinces, enhancing public engagement and local pride. In 2022, UNESCO honoured its 'Five-in-One' World Heritage Education Model as a top ten Global Excellent Recommended Case. New infrastructure, including the Yuedan Rural Road and Danxia Jinjiang Bitao trail, now connects over 70 natural villages (Ng 2015). This integration of education and agriculture has boosted rural opportunities, driven economic growth, and enhanced Mount Danxiashan's tourism offerings, ensuring continued progress.

6.2 Coordinating Protection and Community Development

For over 40 years, Mount Danxiashan has followed a 'joint construction, sharing, and co-governance's philosophy, using global geopark and heritage site practices to ensure conservation and development benefits for all residents. Key initiatives include:

- Revenue Allocation: Ticket income is reinvested—partly subsidizing village resources and partly funding geopark projects.
- Employment: Local villagers receive priority, with 90% of jobs going to rural residents.
- Investment: National, provincial, and municipal funds finance infrastructure like roads, utilities, and waste management.
- Local Services: Community entrepreneurs are encouraged to offer services such as catering and accommodation, while larger projects are managed by major enterprises to ensure sustainable regional growth.

This approach protects the rights of original inhabitants and promotes sustainable development throughout the park.

7 Future

Mount Danxiashan will continue enhancing its geological research, public education, and sustainable resources used by refining management and protection strategies. Key objectives include:

- Enhanced Management and Monitoring: Strengthen ecosystem research, improve flora and fauna monitoring, and advance intelligent management systems.
- Innovative Development Models: Create a 'scientific mountain and smart Danxia' through focused planning, integrating national park practices—linking entrance communities, ecological zones, and science towns while merging production areas with villages and schools. Develop cohesive education centers, homes, paths, and talent systems to support an 'in-park sightseeing, outside-park services' model.

Guided by the principle that 'Lucid waters and lush mountains are invaluable assets', these initiatives highlight that resource conservation is as crucial as economic growth (Xi 2005, 2017). Mount Danxiashan is a national park, global geopark, and World Heritage Site. It implements management practices aimed at balancing nature conservation and human activities, offering benefits to local communities.

References

Chen AZ, Lu YT, Ng Y (2015) The principles of geotourism. Springer, Berlin Heidelberg & Science Press, Springer Geography, Berlin, p 264

Danxiashan Administrative Committee (2011) The master plan for the Danxia mountain scenic area 2011–2025. Danxiashan UNESCO Global Geopark, Shaoguan, China

Danxiashan Administrative Committee (2025) Danxiashan UNESCO Globa; Geopark official web site. https://dxs.sg.gov.cn/en/

Huang J (ed) (1994) Proceedings of the 2nd symposium on Danxia landform and tourism development. Econ Geogr (in Chinese) 14(Supplementary Issue)

National Forestry & Grassland Administration (2022) National park spatial layout plan. Joint publication of the National Forestry and Grassland Administration, the Ministry of Finance, the Ministry of Natural Resources, and the Ministry of Ecology and Environment

Ng Y (2011) Geoparks & geotourism: a management approach to conserve valuable geological heritage in China and the Hong Kong

Special Administrative Region. PhD thesis. School of Geosciences, The University of Sydney, Sydney. pp 126–147
Ng Y (2015) Global geopark development and assessment criteria. Staff Training Note. Danxiashan UNESCO Global Geopark Revalidation Project 2015, Shaoguan, Guangdong
Peng H, Zhang J, Zhou TT (2005a) A discussion on the development of science tourism in Danxia landform areas: a case study of Danxia mountain world geopark (theory and practice of tourism earth science: proceedings of the 11th tourism earth science symposium). China Forestry Publishing House, pp 55–58
Peng H, Zhao F, Zhou TT (2005b) Research methods and applications of tourism demand surveys: a case study of Danxia Mountain world geopark (theory and practice of tourism earth science: proceedings of the 11th tourism earth science symposium). China Forestry Publishing House, pp 317–322
Weng S, Peng H (2014) Tourism development, rights consciousness and the empowerment of Chinese historical village communities. Tour Geogr 16(5):772–784
Xi JP (2005) *'Clear waters and green mountains are as valuable as gold and silver mountains'* - a short commentary from 'Zhijiang Xinyu' on the front page of Zhejiang Daily, 15 August 2005, Zhejiang. https://en.wikipedia.org/wiki/Clear_waters_and_green_mountains. Accessed 12 Jan 2025
Xi JP (2017) Report of the 19th national congress of the communist party of China, Beijing. http://www.xinhuanet.com/english/special/19cpcnc/index.htm. Accessed 12 Jan 2025

Education and Research

Luobin Yan and Young Ng

Abstract

There are many famous mountains in China, such as Mount Everest, known as the 'third pole in the world', Mount Taishan, known as the 'first of the Five Mountains', and Mount Huangshan, 'the most amazing mountain in the world'. Mount Huashan, the Leshan Giant Buddha, which is internationally renowned for its religious and cultural wonders, and Mount Danxiashan are undoubtedly the only famous mountains in China using science to portray as its main external image. The image of Mount Danxiashan as a famous scientific mountain has benefited from the scientific research work carried out by generations of scholars over the past century.

1 The Scientific Research History of Mount Danxiashan

There are many famous mountains in China, such as Mount Everest, known as the 'third pole in the world', Mount Tai, known as the 'first of the Five Mountains', and Mount Huangshan, 'the most amazing mountain in the world'. Mount Huashan, the Leshan Giant Buddha, which is internationally renowned for its religious and cultural wonders, and Mount Danxiashan are undoubtedly the only famous mountains in China using science to portray as its main external image. The image of Mount Danxiashan as a famous scientific mountain has benefited from the scientific research work carried out by generations of scholars over the past century.

L. Yan
Southwest University, Sichuan, China

Y. Ng (✉)
Geotourism Australia, Australian Geoscience Council, Sydney, Australia
e-mail: oz.geotourism@gmail.com

The scientific research history of Mount Danxiashan began with geologist Feng Jinglan's investigation of Mount Danxiashan nearly a century ago. In 1928, during a geological survey in northern Guangdong, Feng named a set of red rock series that constitutes the Mount Danxiashan the 'Danxia Formation' and described the landform characteristics and causes of the development of the 'Danxia Formation' (Feng and Zhu 1928). This is considered the beginning of the study of 'danxia landsacpe', and the word 'danxia' originating from Mount Danxiashan has become an academic term in the field of earth science.

In 1935, Chen Guoda published 'The Red Rock System of Guangdong', which described the special terrain of Mount Danxiashan in detail. Later (Chen 1935). In 1939, Chen officially used the term 'danxia landsacpe' in his research (Ref). Since then, danxia landscape has been used as an academic term in geomorphology in the field of geoscience in China. Afterwards, a group of scholars conducted varying degrees of research and discussion on the stratigraphy, lithology, structure, and landform development of the Danxia Formation in Mount Danxiashan. In 1980, Zeng Zhaoxuan and Professor Huang Shaomin elaborated on the development and geomorphic characteristics of the major danxia landforms in southern China represented by Mount Danxiashan in their book entitled 'China's Physical Geography and Landforms' (Zeng and Huang 1980). In 1982, Huang Jin summarized the basic ways and' slope characteristics of near-horizontal red layer development on slopes under humid climate conditions, mainly based on the understanding gained during the investigation of Mount Danxiashan (Huang 1982). So far, the academic term danxia landscape originating from Mount Danxiashan has been extensively recognized and accepted by the academics within China. This encourages more academic research on danxia landscape which then gradually has been regarded as an independent landscape type and is being treated as a unique research topic for many geomorphologists.

Y. Ng and W. Eder (eds.), *The Redstone Park of China*, Geoheritage, Geoparks and Geotourism,
https://doi.org/10.1007/978-3-031-95603-4_8

Fig. 1 Professor Peng Hua's dedication and tireless efforts to promote danxia landscape within and outside China since the 1980's was extremely important to attract the attentions of not only scholars across the globe but also the public (*Source* Southern Metropolis Daily 20 February 2019)

Before 1991, the research on Mount Danxiashan was in the stage of sporadic, individual studies, mainly focusing on specific problems. In 1991, driven by Huang Jin, Chen Chuankang and others, Mount Danxiashan hosted the first National Academic Symposium on Tourism Development of Danxia Landscape, which resulted in the establishment of a research association to coordinate works from different parties and researchers within China (Huang et al. 1991). Due to the continual efforts to promote research on danxia landscape, a significant number of researchers from across the country started to conduct research on danxia landscape, paving its greater research development in the future (Fig. 1).

Driven by Peng and others, in 2004 Mount Danxiashan had successfully acquired Global Geopark status (Peng et al. 2013; Peng 2020). In 2010, Mount Danxiashan and five areas with distinctive danxia landscape in the humid areas of southern China jointly applied for World Heritage under the name 'Chinese Danxia'. The success of the accreditation has, to a certain extent, promoted the international academic community's understanding of the research results of China's danxia landscape and its acceptance by Chinese scholars. Especially in May 2009, Mount Danxiashan hosted the first International Academic Symposium on Danxia Landscape. Scholars from the fields of geomorphology, geology and geography from 13 countries jointly issued a declaration affirming the achievements and contribution of Chinese scholars in the study of danxia landscape and hence calling for international recognition of danxia landscape. So far, danxia landforms has been studied as an independent landscape type and has attracted the attention of international academia.

After the successful application for World Heritage, the research trend of danxia landscape has changed. The key change is the shift from traditional geomorphological methods to more quantitative research methods. For instance, the research team led by Professor Zhang Ke of Sun Yat-sen University employed modern analytical testing techniques to elucidate the origin of the red rocks in Mount Danxiashan. Studies of honeycomb and the micro-scale landform of danxia landscape have been carried by the research team of Professor Guo Fusheng and Professor Chen Liuqin of East China University of Science and Technology, which uses a series of modern analytical testing methods (Chen et al. 2021, 2022; Zhang et al. 2013, 2024). In addition, Professor Zhu Cheng's team at Nanjing University used experimental geomorphology to study the impact of rock mechanical properties on the development of mesoscale landscape-bedding caves (Zhu et al. 2009). Besides, the research team headed by Professor Liao Wenbo of Sun Yat-sen University's conducted a systematic survey of the biodiversity of Mount Danxiashan and its relationship with the unique landforms (Liao and Fan 2024). They discovered 10 endemic species. With the development of Danxia Mountain tourism, many tourism geography studies using Mount Danxiashan as a case have emerged in recent years.

In the third decade of the 21st century, as Mount Danxiashan was preparing to build a national park, the Guangdong Provincial Geological Survey took the lead in hosting a special geological and landform scientific research project on Mount Danxiashan. The project drew many geological and geomorphological scholars to research Mount Danxiashan.

Research of its stunning landforms now includes evaluating their aesthetic quality.

After the successful application for World Heritage, Mount Danxiashan continued to work hard to expand the international recognition of Danxia landforms. In October 2011, Mount Danxiashan hosted the first meeting of the IAG Red Beds and Danxia Geomorphology Working Group and the second International Academic Symposium on Danxia Landscape. Mount Danxiashan funded Dr. Pan Zhixin to go to St. Louis University in the United States to conduct collaborative research and complete his doctoral thesis 'A Comparative Study of the Development of danxia landforms in Zion National Park in the United States and China's Mount Danxiashan' (Pan and Ren 2020).

Mount Danxiashan is the model site for the theoretical construction of danxia landforms. The understanding of different types of such landforms, their development, and distinctive features originates from numerous studies in Mount Danxiashan. The scientific research history of Mount Danxiashan mirrors the evolution of danxia landform studies. As the naming place of danxia landscape, Mount Danxiashan has played an important role in every important node in the development of the danxia landscape discipline. Scientific research has led to the accumulation of a substantial amount of research materials, making Mount Danxiashan a notable location for scientific study and a base for scientific teaching and research.

2 The Scientific Research Strength of Mount Danxiashan

The scientific research of Mount Danxiashan primarily stems from the dedicated efforts of researchers from esteemed scientific institutions. Several research teams have long been active in Danxia Mountain, including the geology, geomorphology, and ecology team from Sun Yat-sen University, the geology team from East China University of Technology, and the geomorphology team from Southwest University. Recently, as Mount Danxiashan prepares to become a national park, additional teams from the Guangdong Provincial Geological Survey, Shantou University, Guangzhou Institute of Geography, and Hohai University have also contributed to the research efforts.

The most significant scientific research contributions come from Sun Yat-sen University, where four generations of scholars have worked diligently. In 1928, Professor Feng Jinglan and his colleagues named the red strata making up Danxia Mountain and the corresponding strata in northern Guangdong the 'Danxia Formation' (Feng and Zhu 1928). During the 1930s and 1940s, Professor Chen Guoda and others proposed the concept of 'Danxia topography', promoting its study in South China (Chen 1935). From the 1940s to the 1960s, Professors Wu Shangshi and Zeng Zhaoxuan conducted systematic research on red bed landforms and facilitated the widespread dissemination of danxia landforms (Wu and Zeng 1946, 1948). Since the 1980s, Professor Huang Jin has conducted extensive field studies on over 900 Danxia landforms across 28 provinces, advancing the study of danxia landforms nationwide (Huang 1982, 1992, 1994, 1996; Huang et al. 1991).

In the early 21st century, Peng Hua led the effort to apply for Global Geopark status for Mount Danxiashan under the danxia landscape category, which gained acceptance and support from UNESCO and the International Federation of Geosciences (Peng et al. 2004). From 2006 to 2010, Peng Hua successfully managed the joint declaration of representative Danxia landform areas in six southern provinces of China for World Natural Heritage status (Peng 2012a, b). He also organized two international conferences at Mount Danxiashan, continually promoting its global scientific significance.

In recent years, Mount Danxiashan's scientific research capabilities, fields of study, and technological methods have diversified significantly (Fig. 2).

Fig. 2 The Sun Yat-Sen University in Guangdong is one of the earliest universities in China to conduct research on danxia landscape (*Source* Baidu pictures)

3 Challenges in Scientific Research on Mount Danxiashan

Mount Danxiashan, currently in the late prime of the development of danxia landforms, offers the most abundant landscape types, providing opportunities for landform research at different scales. Additionally, Mount Danxiashan has a rich accumulation of basic geological and geomorphological data, making it an excellent case study site for danxia landscape and a potential research base for similar landforms worldwide. However, scientific research on Mount Danxiashan faces several difficulties and challenges, which are not unique to this site but also prevalent across all danxia landform studies.

3.1 Shortage of Research Talent

There is a significant shortage of research talent in the study of Danxia landforms in China. According to Huang Jin, Peng Hua, and Yan Luobin, 1,119 Danxia landforms have been identified in China, distributed across almost all provincial regions except Shanghai, Hong Kong, and Macau (Huang 1996; Peng et al. 2013). Of these, 338 have been developed into scenic spots of various levels. Danxia landforms are mostly found in economically backward hilly and mountainous areas, making their development and utilization crucial for local economic growth and the advancement of science, education, and tourism. Despite the importance of research on Danxia landforms, there is a notable lack of research talent, especially young scholars. Traditional geomorphological research methods, which relied mainly on observation and description, no longer suffice for modern geomorphology that requires quantitative research methods. Therefore, there is an urgent need for young scholars with modern scientific and technological skills to engage in in-depth basic research. High-level researchers are essential to promote the development of Danxia landform disciplines and the internationalization of academic research.

3.2 Insufficient Research Funding

Investment in scientific research on danxia landforms is inadequate. Despite being the most studied danxia landforms in the world, Mount Danxiashan's research funding falls short compared to world-class scientific mountains like the Los Angeles Mountains and Mount Mauna Kea in Hawaii. The National Natural Science Foundation of China (NSFC) plays a crucial role in funding basic research and nurturing outstanding talent. In the past decade, the NSFC has funded 19 Danxiashan related research projects, primarily on ecology, with only eight on geology and geomorphology. Among these, only five projects used Mount Danxiashan as a case study site. The limited funding dampens the enthusiasm of young researchers to focus on danxia landform studies, affecting the growth of young talents and the promotion of professional titles.

3.3 Lack of Unified Scientific Development Plan

The scientific research funding for Mount Danxiashan primarily comes from competitive funding obtained by research teams, leading to fragmented research topics and a lack of sustained motivation. To address this, the Danxia Research Institute was established to develop medium- and long-term plans for the scientific development of Mount Danxiashan and to continuously promote high-quality research outcomes. These challenges have resulted in a scarcity of high-level research results on the geology and geomorphology of Mount Danxiashan. This issue is not limited to Mount Danxiashan but affects all areas with danxia landforms (Fig. 3).

Fig. 3 Research students of Sun Yat-Sen University examining the sandstone and conglomerate in Danxiashan (*Source* Baidu Pictures)

4 Research Project Opportunities in Mount Danxiashan

Through the dedicated efforts of several generations of Chinese geoscientists, a comprehensive understanding of the spatial distribution patterns, conceptual definitions, classification systems, and landscape differentiation of danxia landforms has been established, forming the foundational framework of the subject. Particularly in the past 30 years, scholars like Huang Jin and Peng Hua have conducted nationwide systematic surveys, deepening our understanding of the origins of danxia landforms in China and the factors influencing landscape differentiation (Huang 1982, 1991, 1992, 1994, 1996; Peng 2020). While surface investigations of the landscape in Mount Danxiashan have been thorough, future research will focus on geomorphological processes in specific areas with danxia landforms. Despite progress, the academic foundation of danxia landscape studies remains relatively weak.

At the 2011 Red Beds and Danxia Geomorphology Working Group Meeting of the International Association of Geomorphologists in Danxiashan, 95 scholars from 9 countries issued an initiative to deepen basic research, strengthen core disciplinary issues, and produce high-quality results. Only by enhancing basic research, international cooperation, and the application of advanced technical methods can the geomorphological meaning of danxia landscape achieve international academic recognition comparable to karst landscape.

Research on danxia landscape in Mount Danxiashan spans 90 years. Despite significant progress, qualitative descriptions dominate the study of slope development methods, necessitating stronger quantitative research on theoretical issues. The most notable topographic feature, the steep cliff slope (Red Cliff Danxia) formed by collapse, results from weathering and indentation of the soft rock layer at the cliff's base, leading to the upper rock mass collapsing under gravity. Understanding the development pattern of slopes of danxia landscape is crucial, yet current research remains at a qualitative stage. Experimental geomorphology, engineering geology, numerical simulation, and multidisciplinary approaches hold potential for quantitatively solving these developmental patterns.

Mount Danxiashan is renowned for its danxia landscape', a landscape formed under unique geological and historical conditions. The past century, especially the last decade, has seen advancements in understanding micro-scale landform evolution through quantitative research methods. However, the long-term evolution of danxia landscape remains unexplored. Long-term landform evolution has been challenging in geomorphology, but cosmogenic nuclide techniques offer revolutionary insights. These techniques have great potential to uncover the long-term evolution of danxia landforms and the mechanisms of various external forces.

The evolution mechanism of landforms is a core theoretical issue in danxia geomorphology. Current understanding remains conceptual, lacking quantitative research on long-term slope evolution. Progress in cosmogenic nuclide dating has revolutionized geomorphology research, making it possible to reveal the long-term evolution of danxia landscape. Conducting quantitative research on slope evolution rates using in-situ cosmogenic nuclides is essential for understanding the long-term evolution of danxia landscape and developing a comprehensive disciplinary theory.

Mount Danxiashan offers a treasure trove of research opportunities beyond its ecological, geological, and geomorphological significance. One critical area is environmental conservation. Research here could explore strategies to preserve the region's unique biodiversity while balancing human activity and economic growth. This includes assessing the effectiveness of current conservation methods and proposing new, innovative approaches to protect the fragile ecosystem.

Another significant research avenue is park management and planning. This involves developing comprehensive strategies for managing natural resources, ensuring the sustainable use of the park's assets, and planning for future development. Visitor management is also an essential topic, focusing on ways to enhance visitor experiences while minimizing their impact on the environment. This could include studies on visitor behavior, the carrying capacity of the park, and the development of educational programs to promote responsible tourism.

Rural development is another promising area of research. By examining how tourism and park activities can contribute to the local economy, researchers can propose ways to foster sustainable development that benefits the local community. This ties closely with sustainable development, where the focus is on balancing economic growth with environmental preservation. Studies could explore how to integrate green infrastructure, renewable energy sources, and sustainable practices into the development plans for the region.

Geoscience popularization is a fascinating research topic that involves finding effective ways to educate the public about the park's geological significance. This could include developing educational materials, organizing community outreach programs, and leveraging technology to create engaging and informative content. Lastly, transportation research could examine how to improve access to Mount Danxiashan while adhering to environmentally friendly practices. This might involve studying the impact of current transportation methods on the environment and proposing sustainable alternatives.

Overall, these research opportunities at Mount Danxiashan hold the potential to create a model of sustainable development that not only preserves the natural beauty and ecological significance of the region but also promotes economic growth and community well-being (Fig. 4).

Fig. 4 Mount Danxiashan is often chosen by school teachers for field studies due to its unique geological and geomorphological characteristics, along with its organized and safe study trails (*Source* Baidu Pictures)

5 Funding Sources

Beyond direct financial and human resource support from the geopark, other funding sources are available:

5.1 National Natural Science Foundation of China (NSFC)

The NSFC offers various grants, including the Research Fund for International Scientists (RFIS), which supports international scientists conducting research in China. The RFIS has three sub-types: RFIS-I for young scientists, RFIS-II for excellent young scientists, and RFIS-III for senior scientists (STINT 2020). Funding amounts range from 200,000 to 800,000 RMB per year per project.

5.2 Chinese Academy of Geological Sciences (CAGS)

The CAGS, through the Chinese Geoparks Network (CGN), provides financial support for geoparks. The CGN coordinates national and global geopark activities and assists in their development (NSFC 2023).

5.3 China Geological Survey (CGS)

The CGS supports geopark research and development through various funding programs (Zheng 2020).

5.4 National Forestry and Grassland Administration (NFGA)

The NFGA is responsible for the administration of geoparks in China and provides financial support for their development and management (Zeng 2020).

Additionally, the revised 'Science and Technology Popularization Law of the People's Republic of China' will come into effect on 25 December 2024. The revised law emphasizes that 'science popularization is a public endeavor and a crucial aspect of the construction of socialist material and spiritual civilization'. Science popularization is a long-term national mission, and the country actively promotes the comprehensive integration of science popularization into economic, political, cultural, social, and ecological civilization construction. Such integration will certainly establish a collaborative promotion framework involving the government, society, and the market. Special funding is provided specifically for geoscience education goals and public engagement by the following organizations.

5.5 China Association for the Advancement of Science (CAAS)

The CAAS plays a significant role in science popularization and has been working to improve funding mechanisms for grassroots science popularization activities. The central government has been increasing its investment in science popularization, with national funding reaching a record 21.5 billion yuan ($2.74 billion) in 2023 (ECNS 2025).

5.6 China University of Geosciences (CUG)

The CUG is involved in geoscience popularization efforts and often collaborates with other institutions to promote geoscience education and outreach (IJIRES 2017).

5.7 China Meteorological Administration Training Centre

This center focuses on geoscience education and popularization, providing training and resources to enhance public understanding of geosciences (IJIRES 2017).

5.8 Provincial and Local Governments

Many provincial and local governments in China also provide funding for science popularization initiatives, including geoscience education and outreach programs.

These sources offer a range of funding opportunities for promoting geoscience education and public engagement which can be used for encouraging continual research in Mount Danxiashan.

References

Chen GD (1935) Red rock series of Guangdong. Q Natl Univ Sun Yat Sen (Nat Sci Ed) (in Chinese) 6(4)

Chen, Z & Yang G (2017). Analysis on Geoscience Popularization of Innovation in China basing on functional classification. Int J Innov Res Educ Sci (IJIRES) 4(3):2349–5219. https://www.ijires.org/administrator/components/com_jresearch/files/publications/IJIRES_902_FINAL.pdf?form=MG0AV3. Accessed 19 Jan 2025

Chen X, Chen L, Zhang Y et al (2021) Lithological and environmental controls on large tafoni along conglomerate cliffs in subtropic humid Danxiashan UNESCO Global Geopark. J Mt Sci 18:1131–1143

Chen L, Shi Y, Ng Y et al (2022) Roles of algae in honeycomb weathering in humid climate: an example from Danxiashan UNESCO global geopark of China. Geoheritage 14:120

ECNS (2025) Science popularization funding reaches a record. China Daily, 02 January 2025. https://www.ecns.cn/news/2025-01-02/detail-ihemmuxf3071000.shtml?form=MG0AV3. Accessed 19 Jan 2025

Feng JL, Zhu HS (1928) Geology and mineral resources of Qujiang, Renhua, Shixing and Nanxiong, Guangdong Province. Ann Guangdong Guangxi Geol Surv (in Chinese) 1

Huang J (1982) A basic form of slope development of Danxia Landform. Trop Geogr (in Chinese) 3(2):107

Huang J (ed) (1992) Proceedings of the 1st symposium on Danxia Landform and tourism development. Trop Geomorphol (in Chinese) (Supplementary Issue)

Huang J (ed) (1994) Proceedings of the 2nd symposium on Danxia Landform and tourism development. Econ Geogr (in Chinese) 14(Supplementary Issue)

Huang J (ed) (1996) Proceedings of the 3rd symposium on Danxia landform and tourism development. Econ Geogr (in Chinese) 16(Supplementary Issue)

Huang J et al (1991) A preliminary study of Danxia cave geomorphology. In: Theory and practice of geotourism—collection of geotourism papers, vol II

Liao WB, Fan Q (2024) Geological lithology hypothesis to explain the aggregation of mountain flora. Botanical Society of China. http://www.isenlin.cn/sf_50D4FD3C8CCF446798F06297D22AA69D_209_7E41374880.html. Accessed 01 Nov 2024

NSFC (2023) Research fund for international scientists call for proposals FY 2023. National Natural Science of China. https://www.nsfc.gov.cn/english/site_1/international/D5/2021/04-02/231.html?form=MG0AV3. Accessed 19 Jan 2025

Pan Z, Ren F (2020) Geomorphic features and evolution process of Danxia landforms in Western United States: a case study at Zion National Park, Utah. In: Mountain Research 2020-2, pp 210–221

Peng H (2012a) The world heritage value of China's Danxia and its protection and utilization. Landsc Archit 1

Peng H (2012b) World heritage protection and tourism development. Tour Trib 21(4)

Peng H (2020) China Danxia. Springer Geography, 410p

Peng H, Zhang K, Liu SR (2004) Academic values of Danxiashan Geopark and the significance of the international comparison of Danxia geomorphology. In: Proceedings of the first international conference on geoparks, geological publishing house, pp 123–130

Peng H, Pan ZX, Yan LB, Simonson S (2013) Research on red beds and Danxia landform. Acta Geogr Sin 9:1170–1181

STINT (2020) Public research and innovation funding actors in China. The Swedish Foundation for International Cooperation in Research and Higher Education (STINT) (2025) Stockholm, Sweden. https://www.stint.se/wp-content/uploads/2020/09/23087_STINT_rapport_Public_Research_and_Innovation_Funding_Actors_in_China_webb.pdf?form=MG0AV3. Accessed 19 Jan 2025

Wu SS, Zeng ZX (1946) The red beds in north Kwangtung. Lingnan J (Special issue)

Wu SS, Zeng ZX (1948) Geology and landform of red rock series in north Kwangtung. Geol Assem (in Chinese) 1948(6):13

Zeng ZX, Huang SM (1980) Red bed geomorphology and granite geomorphology. In: Physical geography geomorphology of China (in Chinese). Beijing, Science Press, p 139

Zhang W et al (2013) Morphometric analyses of Danxia landforms in relation to bedrock geology: a case of Mt. Danxia, Guangdong Province, China. Open Geol J 7:54–62

Zhang Z, Chen L, Ng Y et al (2024) Characterization and quantitative assessment of geodiversity in Danxiashan UNESCO Global Geopark of China. Geoheritage 2024–12

Zheng Y (2020) The 2019 Annual Report on the Chinese Geoparks Network. 20 April 2020. Chinese Geoparks Network. https://www.globalgeoparksnetwork.org/sites/default/files/2024-04/Annual-Report-2019-China-1.pdf?form=MG0AV3. Accessed 28 Dec 2024.

Zhu C, Peng H et al (2009) Age and genesis of the Danxia landform on Mount Jianglangshan, Zhejiang Province. J Geogr Sci 19(5):615–630

Geoscience Popularization

Yao Meng and Xin Li

Abstract

In the 21st century, geoparks have emerged globally with the crucial mission of educating the public about Earth's history and the environment. UNESCO Global Geoparks focus on leveraging the geological, ecological, and cultural elements of these parks to conduct various environmental education activities aligned with the United Nations Sustainable Development Goals (SDGs). In concert with the significant efforts of organisations such as ProGEO, IUGS, and IUCN, the geopark initiative enhances public appreciation of geoheritage and promotes greater awareness of environmental protection in the context of climate change. Additionally, the popular science education activities enrich the tourism content of geoparks, enhancing their appeal and promoting local economic development.

1 Introduction

In the 21st century, geoparks have emerged globally with the crucial mission of educating the public about Earth's history and the environment. UNESCO Global Geoparks focus on leveraging the geological, ecological, and cultural elements of these parks to conduct various environmental education activities aligned with the United Nations Sustainable Development Goals (SDGs). In concert with the significant efforts of organisations such as ProGEO, IUGS, and IUCN, the geopark initiative enhances public appreciation of geoheritage and promotes greater awareness of environmental protection in the context of climate change. Additionally, the popular science education activities enrich the tourism content of geoparks, enhancing their appeal and promoting local economic development.

According to the Statutes of the International Geoscience and Geoparks Programme (UNESCO 2015), science popularization activities are a key component in the assessment and evaluation of UNESCO Global Geoparks. Each UNESCO Global Geopark has developed a system of facilities to support these activities, including museums, interpretive signs (hardware), science popularization texts and images (software), as well as management bodies and service personnel.

As a representative of Global Geoparks, Guangdong's famous Mount Danxiashan has fully leveraged its unique resources for over twenty years. With academic support from experts and scholars, and the dissemination efforts of science volunteers, Danxiashan conducts activities based on local resources, collaborates with related science tourism organizations, and guides local community residents to participate in science popularization activities. They have continuously innovated the forms of these activities, gradually establishing the 'Danxiashan Paradigm' for science popularization activities in global geoparks (Fig. 1).

2 Exploring the Science Popularization Model

In response to the intense competition among tourist destinations within China over the past two decades, Mount Danxiashan has developed a unique and effective strategy to attract high-quality tourists. By leveraging its available resources and support, Mount Danxiashan has transitioned from conventional tourism to a more education-based tourism model (Meng et. al 2021; Li and Xie 2023). This developmental shift can be visualized through four distinct stages.

Y. Meng (✉)
China University of Geosciences, Wuhan, China
e-mail: mydili@cug.edu.cn

X. Li
Wuhan Daye Geological Environment Protection Co., Ltd., Wuhan, Hubei Province, China

Y. Ng and W. Eder (eds.), *The Redstone Park of China*, Geoheritage, Geoparks and Geotourism,
https://doi.org/10.1007/978-3-031-95603-4_9

Fig. 1 Mount Danxiashan, one of the world's oldest geoparks, is committed to the pursuit of continual improvement

2.1 Pre-2004 Early Research Stage

During this period, research and exchange activities conducted by various institutes focused on the Danxia landscape, aiming to explore opportunities for geoscience popularization and potential applications in sustainable tourism. This century-long research laid the foundation for the science popularization of Mount Danxiashan. The term 'Danxia landscape,' named after Mount Danxiashan, is a new earth science terminology gradually recognized by Chinese geomorphologists and increasingly accepted by international scholars and researchers. The term first appeared in 1928 when Feng named the red glutenite layer of Danxia landforms observed in Mount Danxiashan and the red beds found in northern Guangdong Province as 'Danxia layer' (Feng 1928). In 1939, Chen of Sun Yat-sen University first proposed the term 'Danxia terrain' (Chen 1935, 1938, 1939). From 1946 to 1948, Wu and Zeng conducted a comprehensive and systematic study of the geology and topography of the red rock series in northern Guangdong, thoroughly examining the sedimentary environment, geological structure, and the relationship between lithology and topography development of Danxia Basin (Wu 1946, 1948, 1957, 1960, 1978, 1980; Zeng 1943). This research laid the groundwork for future geomorphological studies of Mount Danxiashan.

In 1982, Huang proposed the basic slope morphological characteristics and slope development process of Danxia landforms (Huang 1982). This was followed by the First National Symposium on Danxia Landscape Tourism Development held in Mount Danxiashan. In 1991, the Danxia Landscape Tourism Development Research Association was established under the leadership of Huang Jin and Chen Chuankang. Since then, research on Danxia landscape has gained popularity. The definition of Danxia landscape has been extensively discussed, and its recognition as a unique landscape type has gradually been accepted by most scholars in academic community. Research on its unique features, characteristics, and formation has become common topics for researchers and students. The new geomorphic terminology of Danxia landscape has become a special landscape type that deserves the attention of geomorphologists worldwide (Fig. 2).

2.2 Growing Stage—Geoparks as Driver (2004–2013)

The journey began when Danxiashan became one of the first global geoparks in the world. Scientific research and public education activities were carried out extensively, adhering to

Fig. 2 Regular international conferences and seminars for geomorphologists and geoscientists are held to enhance awareness and understanding of Danxia landscape

Global Geoparks standards and focusing on parallel promotion by both academic and tourism sectors. The Global Geoparks standards mandate that scientific research and public education are prioritized, with a 'yellow card' issued to geoparks that emphasize construction over popular science education.

In China, several Global Geoparks received 'yellow cards' from evaluation experts, prompting Danxiashan and other geoparks to take the matter seriously. They proactively organized activities on theme days, such as 'World Earth Day.' Concurrently, the research team led by Professor Peng Hua continued to advance the study and promotion of the Danxia landscape (Peng et al. 2005).

In May 2009, Danxiashan hosted the First International Symposium on the Danxia Landscape, culminating in the Danxia Declaration. Later that year, at the 7th Session of the International Conference on Geomorphology in Melbourne, the Council approved the establishment of a Red Beds and Danxia Geomorphology Working Group. This marked the debut of the Danxia landscape, named after Danxiashan, on the highest international academic stage (IAG 2009).

The inscription of China Danxia on the World Heritage List in 2010 further solidified the international recognition of the Danxia landscape (WHC 2010). To promote the Danxia landscape, Danxiashan launched innovative popular science education activities under the 'China Danxia' initiative in colleges and universities.

2.3 Development Stage 1: Popularization (2014–2019)

During this phase, Danxiashan organized and initiated the 'Danxiashan Science Volunteer Training Camp', focusing on exploring scientific tourism based on the scientific essence of Danxiashan, with the aim of establishing it as a 'scientific famous mountain'. In 2014, Danxiashan Global Geopark collaborated with third-party institutions to pioneer a new model of science popularization education through the volunteer training camp (He et al. 2022; Li and Xie 2023; Meng 2021). The initiative aimed to promote the dissemination of Earth science knowledge and the value of Global Geoparks.

The activities not only enhanced public scientific literacy and the park's influence in science popularization but also propelled the development of Danxiashan Global Geopark's science tourism endeavors. Since its inception in 2014, the training camp has continuously iterated and upgraded, cultivating over 600 science volunteers up to date, making it a model for advancing science education in geoparks.

2.4 Development Stage 2: New Concepts (Since 2019)

Danxiashan is the first town in Guangdong Province to launch its knowledge-based science popularization and tourism programs. Since 2019, local communities have been encouraged to actively participate in these programs to foster tourism activities, bringing direct benefits to the local people (Meng 2021). Mount Danxiashan successfully established one of the UNESCO Global Geopark training bases in 2019 and organized the '2019 Danxiashan Field Exchange on Popular Science Education of Chinese UNESCO Global Geoparks', which set an example and model for other geoparks both within China and abroad. In recent years, Danxiashan Geopark has held over 300 popular science education events, including essay contests, nature notes contests, starry sky conferences, volunteer training camps, and public lectures for schools and the community. These activities aim at enhancing the branding of Mount Danxiashan as a mountain for science and learning.

Mount Danxiashan has also established one of China's first National Geo-culture Villages, which integrates culture and rural tourism with geology, ecology, rural environment and tourism. The local geological, ecological, and cultural resources are fully utilized, and numerous transformation projects have been carried out to simplify scientific and technical concepts into understandable expressions for applying in science popularization and tourism activities. The Geo-culture Village concept emphasizes unique village characteristics, transforming them into tourist attractions.

3 Representative Activities

As a World Heritage Site and one of the world's first global geoparks, Mount Danxiashan has launched the above-mentioned innovative activities such as Danxiashan Popular Science Volunteer Training Camp, Danxiashan outreach and public lectures and talks for the community and schools. The geopark also encourages and promotes the involvement of community in science popularization tourism, transforming it into a town of popular science tourism (Table 1).

3.1 Danxiashan's Popular Science Volunteer Training Camp

In 2014, Danxiashan UNESCO Global Geopark pioneered the popular science volunteer training camp. So far, 334 science volunteers have been trained (Table 2). The data does not

Table 1 Four representative popular science education activities of Danxiashan Global Geopark

Activity	Purpose	Participants	Time/Place	Contents	Duration	Rules	Frequency
Danxiashan Popular Science Volunteer Training Camp	To foster popular science education talents and popularize the science of Danxiashan	Young fans represented by university and college students; experts	Danxiashan and cities where the volunteers live	On-site training on Danxia landform, ecology and culture of Danxiashan; popular science promotion activities on the knowledge system related to Danxiashan in the cities where the volunteers live	On-site training in Danxiashan for 5–7 days; the service period of the popular science volunteers is not less than 1 year	25–30 people	1 or 2 courses per year
Mount Danxiashan: Outreach and Lectures	To disseminate the popular science knowledge of Danxiashan and promote public awareness on nature and environment protection	School and university students; popular science instructors of Danxiashan	Schools and universities around China	Sharing the knowledge of geology, ecology and culture of Danxiashan in the form of popular science lectures and donating popular science books to schools and universities around China	1–2 h	No limit	Not less than 200 lectures per year
China Danxia Popular Science lectures on Campus	To promote the scientific, ecological and cultural values of Danxia landform	University and college students; experts	Major universities and colleges	Sharing the scientific value of Danxia landform and Danxiashan in the form of popular science lectures	2–3 h	No limit	1–3 lectures per year
Building Danxiashan Popular Science Tourism Town	To promote popular science tourism in Danxiashan and sustainable development of the region by means of popular science education activities	Tourists, community residents, popular science volunteers	Danxiashan	Geology, ecology, culture and etc. of Danxiashan and surrounding areas	No limit	No limit	No limit

Table 2 Recruitment status of Danxiashan's popular science volunteer training camp

Year	报名人数(人) Number of applications	提交有效申请(份) Number of valid applications	最终人员数量(人) Number of enrolled campers
2014	389	/	30
2015	600	/	30
2016	460	194	28
2017	2,463	1,200	30
2018	1,561	1,204	30
2019	2,340	1,950	30
2020	2,024	1,624	66
2021	1,192	1,052	60
2022	835	316	30

include the number of local volunteers in Mount Danxiashan, and it is as of December 31, 2022).

A course of Danxiashan's popular science volunteer training camp consists of three phases, namely the preparation phase, the implementation phase, and the post-course volunteer service phase. During the one-month preparation phase, recruitment information is posted online. Applicants fill in a form that gathers their basic information, expertise, and special skills for quick and accurate candidate screening. Telephone or online interviews are conducted to evaluate and select the final campers. During the implementation

Fig. 3 Regular training activities are available for tourist guides, teachers and volunteers

phase (lasting about 5–7 days), the selected campers gather in Mount Danxiashan to participate in the popular science volunteer training camp. Different activities are arranged every day, with experts of Danxiashan leading the field trips during the day and exchanging and learning in groups in the evening. Campers who successfully pass the examination will be designated as official popular science volunteers of Mount Danxiashan. During the volunteer service phase (lasting one year), volunteers from different places and occupations return to their workplaces or schools to carry out volunteer service activities. And Mount Danxiashan provides material, technical, personnel and welfare support for the volunteers' service activities, and outstanding volunteers are given praise.

Danxiashan's popular science volunteer training camp has expanded the influence of Mount Danxiashan in China through the Internet platform, while addressing the geopark's shortage of popular science education talents. This pro-environmental behavior (Chen 1935) strengthens the volunteers' sense of fulfillment and pride, which is one of the important motivations for their popular science education activities. In 2015, Danxiashan's popular science volunteer training camp was included in China's popular science education demonstration activities. And other geoparks visited Mount Danxiashan to learn from its experience in organizing popular science education activities, which contributed to the cultivation of popular science education talents in Chinese geoparks (Fig. 3).

3.2 'Qimei Tiancheng Mount Danxiashan' Train-The-Trainer Program

'Danxiashan: Masterpiece of the Gods' is one of the significant popular science books of the Danxiashan UNESCO Global Geopark. To provide free access and raise awareness about Mount Danxiashan, a book drifting campaign was launched in November 2018 for students across schools and universities nationwide.

The campaign began with online training for popular science lecturers. Experts from Mount Danxiashan trained enthusiasts on the park's knowledge to ensure scientific accuracy, transforming these enthusiasts into lecturers. The second stage involved offline activities where these lecturers conducted popular science talks and distributed copies of 'Danxiashan: Masterpiece of the Gods' in schools and universities.

In 2019, the campaign held 209 lectures across 173 schools and universities in 39 cities within 18 provinces in China, attracting an average audience of 62.5 people per lecture. Through websites, print, and social media, the campaign reached millions, marking it as the largest popular science education initiative in the history of China's geoparks (Fig. 4).

3.3 School Program

In 2009, Mount Danxiashan launched the China Danxia Popular Science Lectures on Campus Campaign in higher education institutions, aiming to make high-level talents aware of the scientific significance, aesthetic value, and research progress of the Danxia landscape. The campaign features popular science lectures by leading experts from Mount Danxiashan, including Professor Peng Hua, Professor Su Dechen, and Professor Guo Fusheng.

The lectures cover various topics such as the theoretical knowledge and research history of the Danxia landform, the outstanding universal value of the China Danxia World Heritage, and the development and utilization of the Mount Danxiashan UNESCO Global Geopark. Each lecture lasts 2–3 h and consists of four key sessions: video promotion of Mount Danxiashan, a popular science lecture, Q&A interaction, and an on-site lucky draw. Each event typically attracts 200–300 participants, with 1–3 events held annually.

The campaign is publicized through online previews, posters, and WeChat promotion. To date, more than 30 science lectures have been conducted in 27 universities, including prestigious institutions like Peking University and China University of Geosciences (Fig. 5).

3.4 Danxiashan Science Popularization Town

The popular science education activities at Mount Danxiashan have significantly contributed to the rise of popular science tourism. In 2017, the 'Danxiashan Nature School' was established to promote sustainable community development, offering a venue for training and skill enhancement for community involvement in science tourism.

In 2019, the geopark introduced the innovative concept of a 'popular science town' to brand and manage local popular science tourism products. This initiative has led to the development of a comprehensive system integrating outdoor expeditions, popular science venues, and classrooms. Through training and exchange activities, such as local instructor and volunteer training camps, Mount Danxiashan has cultivated many popular science instructors and guides.

It has also developed thematically popular science education villages in the surrounding areas, transforming them into specialized hubs like astronomy village, geography village, gardening village, painting village, and fruit village. Each village features unique industries, popular science display centers, instructor teams, distinctive products, and high-quality services, paving the way for innovative rural revitalization.

By the end of 2020, Danxiashan Popular Science Town boasted 27 themed popular science classrooms, 9 candidate classrooms, 5,000 square meters of various education venues, 9 education routes, and over 200 thematic courses covering astronomy, geology, flora and fauna, meteorology, outdoor sports, Zen, Hakka culture, handicrafts, and farming. Annually, Danxiashan Popular Science Town receives 3 million visitors, with approximately 400,000 participating in popular science courses.

Popular science education activities are seamlessly integrated with agriculture, manufacturing, and service industries, combining ecological and cultural resources. This synergy between nature education and the rural economy drives the development of agriculture and handicraft manufacturing. Country fairs and live webcasts are organized to showcase and sell specialty products.

An accommodation service system featuring brand hotels and unique bed and breakfasts (B&Bs) has been established, facilitating their transformation into a cluster of popular science bases. This initiative has benefited nearly 30 local B&Bs (Fig. 6).

4 The Significance of Science Popularization Activities at Mount Danxiashan

Mount Danxiashan's science popularization activities offer enriching experiences in environmental education, outdoor exploration, and cultural heritage. These events are especially encouraged for the community, especially the young people and students to participate and engage with.

4.1 Shifting Traditional Tourism to Knowledge-Based Tourism

Mount Danxiashan, like other traditional scenic areas, follows the tourism destination life cycle theory. As shown in Fig. 1, before 2004, Mount Danxiashan had fewer than 1 million tourists annually, with only modest growth. Scholars agreed that the reason was the insufficient direct communication of the tourism image of Mount Danxiashan, but Danxia landscape had great 'market potential in popular science and adventure tourism' (Bao et al.1995; Peng et al. 2005). They suggested focusing on the Danxia landscape to create a unique tourism image. Accordingly, Mount Danxiashan started the work of creating a world-class brand, and it became

Fig. 4 Prioritizing kids as the most important targets for science popularization is importance for the future development of the geopark

a Global Geopark in 2004 and a Natural World Heritage Site in 2010. From 2004 to 2011, the number of visitors to Mount Danxiashan surged from 734,000 in 2003 to 3,503,000 in 2011, reflecting an impressive average annual growth rate of 21.57%. This rapid increase highlighted the emergence of its world-class brand effect. However, starting in 2012, the number of tourists began to decline, falling below 3 million by 2014.

The primary reason for the decline was the limited recognition of Mount Danxiashan among the Chinese people, and its image positioning was not accurately communicated. Even employees within the Scenic Area had limited understanding of its tourism image. Despite being the namesake and a quintessential representative of the Danxia landscape, Mount Danxiashan has suffered from low public awareness and social engagement (He et al. 2022) (Table 3 and Fig. 7).

To boost tourism, Mount Danxiashan launched the 'Popular Science Volunteer Training Camp' in 2014, focusing on its scientific significance to develop a clear brand image. The tourism development strategy of the geopark has gradually shifted towards popular science tourism. Since 2014, Mount Danxiashan has steadily developed this approach. By 2017, the decline in visitor numbers was successfully reversed, highlighting the positive impact of popular science tourism.

As the namesake of the Danxia landscape, it is a natural choice for Mount Danxiashan to focus on the scientific, typical, and representative aspects of the Danxia landscape. The geopark is transitioning from a traditional tourist spot to a science tourism destination, recognized as a site for scientific interest and study.

4.2 Improving the Income and Quality of Life of Local Communities

Mount Danxiashan, located in Danxia District, Renhua County, has experienced significant economic growth in its rural communities. According to the published per capita annual disposable income of rural residents (Fig. 8), before 2015, the income levels in Danxia District were nearly equivalent to those in Renhua County. However, after 2015, the per capita annual disposable income in Danxia District began to surpass that of Renhua County, reaching over CNY 20,000 by 2019.

This improvement is closely linked to the geopark's strategy of community participation in popular science tourism. Since 2015, Mount Danxiashan has encouraged residents to engage in popular science tourism by providing services such as accommodation, catering, transportation, tour guiding, and ticket discounts. Additionally, themed science activities have been developed that leverage Danxiashan's ecological assets to enhance the well-being of the community.

4.3 Cultivating Talents for Popular Science Tourism

Leveraging the advantages of innovative popular science education in Mount Danxiashan, numerous professionals have been cultivated for the development of the popular science tourism industry. Influenced by the success of Mount Danxiashan in popular science tourism, many volunteers have transitioned into roles as planners, organizers, workers, research scholars, and disseminators of ecological tourism. This community participation has nurtured many local 'talents', allowing residents to shift from traditional agricultural work to roles in knowledge dissemination.

As one of the first UNESCO Global Geopark training bases, Mount Danxiashan has become a learning destination for UNESCO Global Geoparks at home and abroad, sharing its advanced experience in organizing popular science education activities. The geopark has assisted many other global geoparks in enhancing their popular science education in line with their unique conditions.

Fig. 5 Outreach activities to the universities, schools and communities introducing Mount Danxiashan

As a leader in popular science tourism within China's nature reserves, Mount Danxiashan has cultivated a talent pool with rich experience in popular science tourism. It has also trained and dispatched popular science tourism managers to various nature reserves. To date, over 100 nature reserves across the country have studied and adopted the Mount Danxiashan model, establishing their own volunteer teams to enhance visibility and influence. This approach has proven to be a successful promotional strategy and a significant theme in popular science education activities.

4.4 Enhancing the Market Recognition of Mount Danxiashan's Brand as a 'Famous Mountain of Science'

In recent years, Mount Danxiashan has significantly deepened its scientific research and popular science education efforts, transforming into a field practice base for 22 universities both domestically and internationally. A series of popular science education activities have effectively increased the geopark's popularity and boosted tourism.

Designated as a 'UNESCO Global Geoparks Training Base', 'National Popular Science Education Base', 'Popular Science Education Base of Land and Resources', 'National Research and Practice Education Base for School Students', 'Popular Science Education Base of Land and Resources

Fig. 6 Popularising science activities generate tremendous interests of parents who encourage their children to participate in activities organized by the geopark

of Guangdong Province', 'Nature School of Guangdong Province', and 'the first popular science town in Guangdong Province', Mount Danxiashan has gradually solidified its brand as the 'Famous Mountain of Science'.

5 Conclusions

Conducting popular science education activities is one of the key responsibilities of UNESCO Global Geoparks. These activities not only promote the development of popular science tourism but also bring significant benefits to community residents. As a pioneer in popular science education within geoparks, Mount Danxiashan has built a diverse team of experts and established a robust system of popular science education talents. They have actively conducted

Table 3 Some popular science courses of Danxiashan Global Geopark

No	Courses								Peripheral sites
	Geology	Biology	Ecology	Astronomy	Meteorology and hydrology	Handcraft	Culture	Country life	
1	Formation process of danxia landscape	Animal and plant resources of Mount Danxiashan	Ecological diversity	Astronomy around you	Meteorological observation	Ormosia picking and knitting	Cultural tourism resources of Mount Danxiashan	Making Hakkas snacks	Meiguan Ancient Road
2	Danxia landscape	Butterflies	Biodiversity of Mount Danxiashan	Theory of astronomical telescope	Nephogram and Typhoon	Plant stamping and collage	Rubbing the inscriptions on cliffs	Making bamboo-tube rice	Maba Man Site
3	Experiment on the Deposition of danxia landscape	Animals' homes	Survival Wisdom of plants	Disassembly and assembly of astronomical telescope	Thunder and lightning warning and protection	Releasing fireflies	Niubi ancient village	Making traditional Tofu with stone mill	Fankou Mine
4	Gifts from 100 million years ago: fossils	Friends of ant lion	Observing insects at night	Making a simple telescope	Judging the weather by reading the clouds	Hand-painting bamboo pipe	夏 Xiafu ancient village	Making characteristic Hakkas snacks	Dabaoshan
5	Evolution model of danxia landscape	Tricks of plants	Eco-bottle activity design	Observing stars	Drinking water	Rock painting	Shitang ancient village	Catching fish	Wumazhai
6	Demonstration of exogenic process	The life of Shatian pomelo	Observing birds in the morning	Deep space photography	Water processing	Flower arrangement	Experiencing traditional tea culture	Planting dendrobe by imitating the wild environment	Zhuji ancient alley
7	Measuring the hardness of rocks	Bamboo arts	Tree-planting day	Star sharing club	Source of water pollution	Mangonel of ancient Rome	Making green mandarin Pu-erh tea	Crop transplanting and harvesting	Ruyuan Canyon
8	Measuring the density of rocks	Plant specimen	Making tea		Tension of water	Handmade soap	Brewing experience	Crop pollination and bee keeping	Chebaling
9	Experiment on acid and alkaline property of soil	Age of plants			Water distribution	Knowing plants and leaf rubbing	Traditional folk custom—Story playing	Growing mushroom	Crocodile reserve
10	Danxiashan and the nomination of China Danxia World Heritage				Meaning of water	Rattan plaiting		Growing lucid Ganoderma	Renhua Meteorology Museum
11	Plant fossils				Water detection experiment	Making sachet			Nanling Ecology and Meteorology Center
12					Water perception and experiment	Baking			
13						Making bookmarks with plants			
14						Natural nameplates			

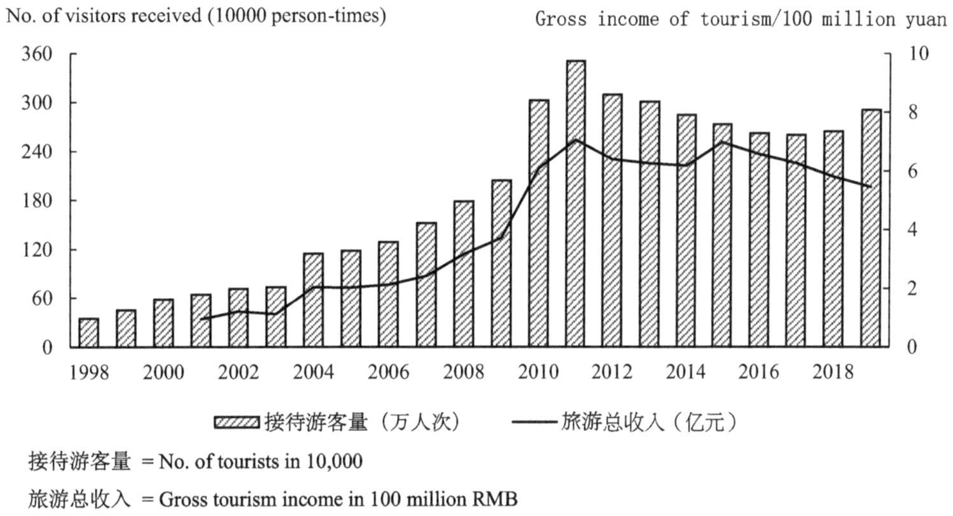

Fig. 7 Visitor number and gross income of Mount Danxiashan (1998–2019)

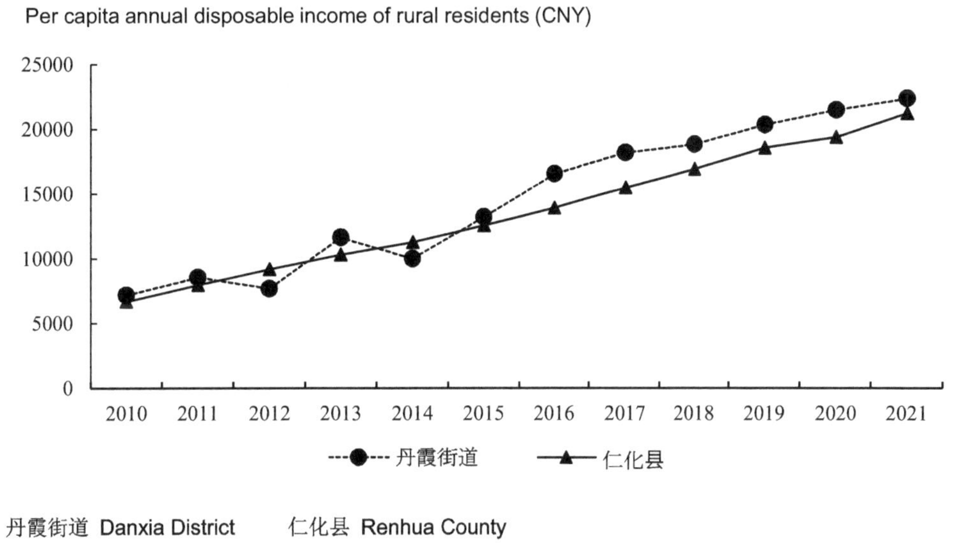

Fig. 8 Comparison of the per capita annual disposable income of rural residents in Danxia District and Renhua County

scientific research, utilized local resources and characteristics, established a comprehensive system for scientific research and knowledge dissemination, and transformed research findings into popular science education materials, continually innovating the content of their educational activities.

By leveraging these popular science education activities, Mount Danxiashan has pioneered brand-oriented development in popular science tourism, creating a multifaceted system of tourism products and services that emphasizes community involvement. This approach has culminated in the formation of the 'Danxiashan Paradigm', characterized by its diverse talent system, promotion of scientific research, innovative popular science education activities, development of a popular science tourism brand, encouragement of community participation, and the industrialization of popular science education.

Furthermore, the geopark has strengthened interdisciplinary, cross-system, and cross-border cooperation and exchange to share authoritative and high-quality popular science resources. This effort has successfully introduced Mount Danxiashan's brand as the 'famous mountain of science' to the world.

References

Bao JG, Peng H (1995) A study on the expansion development of tourist resorts - taking Yangyuan section, Danxia scenic spot as an example. Sci Geogr Sin 15(1):63–70, 100

Chen GD (1935) Red rock series of Guangdong. Q Natl Univ Sun Yat Sen (Nat Sci Ed) (in Chinese) 6(4)

Chen GD (1938) On the subdivisions of the red beds of south-eastern China. Bull Geol Soc China 10(8)

Chen, G. D., Liu, H. S. (1939), Geology of Gongshui valley, Jiangxi. Geological Assembly of Jiangxi (in Chinese), (2).

Feng JL, Zhu HS (1928) Geology and mineral resources of Qujiang, Renhua, Shixing and Nanxiong, Guangdong Province. Ann Guangdong Guangxi Geol Surv (in Chinese) 1

He J, Cai XM, Li GQ et al (2022) Volunteering and pro-environmental behavior: the relationships of meaningfulness and emotions in protected areas. J Sustain Tour 1–18
Huang J (1982) A basic form of slope development of Danxia Landform. Trop Geogr (in Chinese) 3(2):107
IAG (2009) Red beds and Danxia Geomorphology Working Group. International Association of Gemorphologists. https://www.geomorph.org/danxia-geomorphology-working-group/. Accessed 01 Jan 2025
Li GQ, Xie J, Liu J (2023) Study on the mechanism of intellectual elites' development in the knowledge-oriented community tourism. Tour Trib 38(01):109–121
Meng Y, Chen F, Li X et al (2021) Exploration and practice of popular science education through socialized participation in Danxiashan. Nat Prot Areas 1(04):65–71
Peng H, Zhao F, Zhou TT (2005) The survey and analysis of tourism demand in tourism generating markets- a case study of Danxiashan Mountain. J Guilin Inst Tour 01:49–53, 63
UNESCO. Statutes of the International Geoscience and Geoparks Programme (2015) [OL]. https://unesdoc.unesco.org/ark:/48223/pf0000260675. Accessed 02 Jan 2025.
WHC (2010) China Danxia, World Heritage Convention. https://whc.unesco.org/en/list/1335/. Accessed 17 Jan 2025
Wu SS, Zeng ZX (1946) The red beds in north Kwangtung. Lingnan J (special issue)
Wu SS, Zeng ZX (1948) Geology and landform of red rock series in north Kwangtung. Geol Assem (in Chinese) 6:13
Zeng ZX (1943), Preliminary discussion on the landform in thickly bedded red sandstone region in southern Renhua. In: Geographic collection of National University of Sun Yat-Sen (in Chinese), vol 12, p 19
Zeng ZX, Li GZ (1957) Characteristics of South China sandstone landform. Geogr Knowl (in Chinese) 397
Zeng ZX (1960) Rock Geomorphology (in Chinese). Geological Publishing House, Beijing, p 45
Zeng ZX, Huang SM (1978) Red bed geomorphology in southern China. South China Teach Coll J (Nat Sci Ed) (in Chinese) 12
Zeng ZX, Huang SM (1980) Red bed geomorphology and granite geomorphology. In: Physical geography · geomorphology of China (in Chinese). Science Press, Beijing, p 139

Correction to: Geomorphology

Ke Zhang

Correction to:
Chapter 4 in: Y. Ng and W. Eder (eds.), *The Redstone Park of China*, Geoheritage, Geoparks and Geotourism, https://doi.org/10.1007/978-3-031-95603-4_4

In the original version of the book, the following belated correction has been incorporated: In Chapter 4, incorrect figure (figure 22) was updated which has now been replaced with a similar image. The book has been updated with the changes.

Fig. 22 Various Danxia peak assembly in Mount Danxiashan

The updated version of this chapter can be found at https://doi.org/10.1007/978-3-031-95603-4_4

Y. Ng and W. Eder (eds.), *The Redstone Park of China*, Geoheritage, Geoparks and Geotourism,
https://doi.org/10.1007/978-3-031-95603-4_10

Afterword

Mount Danxiashan, with its mesmerizing red stone formations and ever-evolving landscapes, stands as a striking testament to the awe-inspiring power of natural forces that have shaped our planet over millions of years. As one of the earliest geoparks established worldwide in 2004, Mount Danxiashan has navigated numerous challenges along its journey, emerging stronger and more significant with each passing year. To honour its 20th anniversary, dedicated scholars, conservationists, and park managers who have been integral to its development have joined forces to document their rich experiences and insights in this volume, offering it as a gift of knowledge to readers both within China and across the globe. This book aspires to take readers on an in-depth exploration of this extraordinary UNESCO Global Geopark, shedding light on its remarkable geological history, cultural heritage, ecological value, management successes, and ongoing research and educational initiatives.

Across these chapters, we have unravelled the intricate geological processes that have crafted the enchanting danxia landscape. From tectonic uplift to relentless weathering and erosion, these forces have chiselled a terrain of unparalleled beauty and complexity. The striking red hues and dramatic rock formations not only captivate the senses but also unlock profound insights into the region's geological evolution and history.

The ecological diversity of Mount Danxiashan has emerged as a cornerstone of its global significance. As thoroughly explored in this book, the geopark serves as a sanctuary for rare and unique plant and animal species. The tireless dedication of researchers and conservationists has led to groundbreaking discoveries, continually expanding our understanding of its intricate ecosystem. This wealth of knowledge reinforces Mount Danxiashan's status as a vital natural reserve of global importance.

Equally compelling is the cultural richness of Mount Danxiashan, where ancient temples, shrines, and historical artifacts coexist harmoniously with its stunning natural landscape. This seamless blend of nature and culture creates a site of profound historical, spiritual, and artistic significance, inspiring visitors from every corner of the globe. It is a place where human heritage and natural beauty converge, enriching the lives of those fortunate enough to experience it.

The management and development of Mount Danxiashan as a geopark have been pivotal in safeguarding its natural and cultural treasures. Over the years, the unrelenting commitment of park managers and educators has ensured the park's sustainable growth while promoting public awareness and engagement. Through carefully balanced policies, the geopark not only preserves its authenticity and integrity but also fosters sustainable tourism, allowing future generations to learn from and marvel at this natural wonder.

In essence, Mount Danxiashan transcends its role as a geological marvel; it symbolizes the harmonious coexistence of humanity and nature. The insights, experiences, and expertise shared in this volume offer a holistic perspective on what makes this geopark uniquely exceptional. From its majestic landscapes to its role as a beacon of sustainable conservation, Mount Danxiashan embodies the ideals of UNESCO Global Geoparks.

Mount Danxiashan's breathtaking terrain, unparalleled ecological diversity, rich cultural heritage, and ongoing conservation efforts exemplify the very essence of what it means to cherish and protect our planet's most extraordinary treasures. It is our hope that this book has not only informed readers but has also inspired them to champion the preservation of such natural and cultural wonders, ensuring their survival for generations to come. As Mount Danxiashan continues to evolve, it remains a shining example of what humanity can achieve when we work in unison to protect and celebrate the beauty of the natural world.

Y. Ng and W. Eder (eds.), *The Redstone Park of China*, Geoheritage, Geoparks and Geotourism,
https://doi.org/10.1007/978-3-031-95603-4

MIX
Papier aus verantwortungsvollen Quellen
Paper from responsible sources
FSC® C105338
FSC www.fsc.org

If you have any concerns about our products,
you can contact us on
ProductSafety@springernature.com

In case Publisher is established outside the EU,
the EU authorized representative is:
Springer Nature Customer Service Center GmbH
Europaplatz 3, 69115 Heidelberg, Germany

Printed by Libri Plureos GmbH
in Hamburg, Germany